Lecture Notes on the Lambda Calculus

Peter Selinger

Department of Mathematics and Statistics
Dalhousie University, Halifax, Canada

ISBN: 978-0-359-15885-0

This printing: revision 622cc94 of 2023/02/17.

Contents

1 Introduction

1.1 Extensional vs. intensional view of functions

What is a function? In modern mathematics, the prevalent notion is that of "functions as graphs": each function f has a fixed domain X and codomain Y, and a function $f : X \to Y$ is a set of pairs $f \subseteq X \times Y$ such that for each $x \in X$, there exists exactly one $y \in Y$ such that $(x, y) \in f$. Two functions $f, g : X \to Y$ are considered equal if they yield the same output on each input, i.e., $f(x) = g(x)$ for all $x \in X$. This is called the *extensional* view of functions, because it specifies that the only thing observable about a function is how it maps inputs to outputs.

However, before the 20th century, functions were rarely looked at in this way. An older notion of functions is that of "functions as rules". In this view, to give a function means to give a rule for how the function is to be calculated. Often, such a rule can be given by a formula, for instance, the familiar $f(x) = x^2$ or $g(x) = \sin(e^x)$ from calculus. As before, two functions are *extensionally* equal if they have the same input-output behavior; but now we can also speak of another notion of equality: two functions are *intensionally*[1] equal if they are given by (essentially) the same formula.

When we think of functions as given by formulas, it is not always necessary to know the domain and codomain of a function. Consider for instance the function $f(x) = x$. This is, of course, the identity function. We may regard it as a function $f : X \to X$ for *any* set X.

In most of mathematics, the "functions as graphs" paradigm is the most elegant and appropriate way of dealing with functions. Graphs define a more general class of functions, because it includes functions that are not necessarily given by a rule. Thus, when we prove a mathematical statement such as "any differentiable function is continuous", we really mean this is true for *all* functions (in the mathematical sense), not just those functions for which a rule can be given.

On the other hand, in computer science, the "functions as rules" paradigm is often more appropriate. Think of a computer program as defining a function that maps input to output. Most computer programmers (and users) do not only care about the extensional behavior of a program (which inputs are mapped to which outputs), but also about *how* the output is calculated: How much time does it take? How much memory and disk space is used in the process? How much communication bandwidth is used? These are intensional questions having to do with the particular way in which a function was defined.

1.2 The lambda calculus

The lambda calculus is a theory of *functions as formulas*. It is a system for manipulating functions as *expressions*.

---

[1]Note that this word is intentionally spelled "intensionally".

Let us begin by looking at another well-known language of expressions, namely arithmetic. Arithmetic expressions are made up from variables ($x, y, z \ldots$), numbers ($1, 2, 3, \ldots$), and operators ("+", "−", "×" etc.). An expression such as $x+y$ stands for the *result* of an addition (as opposed to an *instruction* to add, or the *statement* that something is being added). The great advantage of this language is that expressions can be nested without any need to mention the intermediate results explicitly. So for instance, we write

$$A = (x + y) \times z^2,$$

and not

$$\text{let } w = x + y, \text{ then let } u = z^2, \text{ then let } A = w \times u.$$

The latter notation would be tiring and cumbersome to manipulate.

The lambda calculus extends the idea of an expression language to include functions. Where we normally write

$$\text{Let } f \text{ be the function } x \mapsto x^2. \text{ Then consider } A = f(5),$$

in the lambda calculus we just write

$$A = (\lambda x.x^2)(5).$$

The expression $\lambda x.x^2$ stands for the function that maps x to x^2 (as opposed to the *statement* that x is being mapped to x^2). As in arithmetic, we use parentheses to group terms.

It is understood that the variable x is a *local* variable in the term $\lambda x.x^2$. Thus, it does not make any difference if we write $\lambda y.y^2$ instead. A local variable is also called a *bound* variable.

One advantage of the lambda notation is that it allows us to easily talk about *higher-order* functions, i.e., functions whose inputs and/or outputs are themselves functions. An example is the operation $f \mapsto f \circ f$ in mathematics, which takes a function f and maps it to $f \circ f$, the composition of f with itself. In the lambda calculus, $f \circ f$ is written as

$$\lambda x.f(f(x)),$$

and the operation that maps f to $f \circ f$ is written as

$$\lambda f.\lambda x.f(f(x)).$$

The evaluation of higher-order functions can get somewhat complex; as an example, consider the following expression:

$$((\lambda f.\lambda x.f(f(x)))(\lambda y.y^2))\ (5)$$

Convince yourself that this evaluates to 625. Another example is given in the following exercise:

Exercise 1. Evaluate the lambda-expression

$$\Big(\big(\,(\lambda f.\lambda x.f(f(f(x))))\,(\lambda g.\lambda y.g(g(y)))\,\big)(\lambda z.z+1)\Big)(0).$$

We will soon introduce some conventions for reducing the number of parentheses in such expressions.

1.3 Untyped vs. typed lambda-calculi

We have already mentioned that, when considering "functions as rules", it is not always necessary to know the domain and codomain of a function ahead of time. The simplest example is the identity function $f = \lambda x.x$, which can have any set X as its domain and codomain, as long as domain and codomain are equal. We say that f has the *type* $X \to X$. Another example is the function $g = \lambda f.\lambda x.f(f(x))$ that we encountered above. One can check that g maps any function $f : X \to X$ to a function $g(f) : X \to X$. In this case, we say that the type of g is

$$(X \to X) \to (X \to X).$$

By being flexible about domains and codomains, we are able to manipulate functions in ways that would not be possible in ordinary mathematics. For instance, if $f = \lambda x.x$ is the identity function, then we have $f(x) = x$ for *any* x. In particular, we can take $x = f$, and we get

$$f(f) = (\lambda x.x)(f) = f.$$

Note that the equation $f(f) = f$ never makes sense in ordinary mathematics, since it is not possible (for set-theoretic reasons) for a function to be included in its own domain.

As another example, let $\omega = \lambda x.x(x)$.

Exercise 2. What is $\omega(\omega)$?

We have several options regarding types in the lambda calculus.

- *Untyped lambda calculus.* In the untyped lambda calculus, we never specify the type of any expression. Thus we never specify the domain or codomain of any function. This gives us maximal flexibility. It is also very unsafe, because we might run into situations where we try to apply a function to an argument that it does not understand.

- *Simply-typed lambda calculus.* In the simply-typed lambda calculus, we always completely specify the type of every expression. This is very similar to the situation in set theory. We never allow the application of a function to an argument unless the type of the argument is the same as the domain of the function. Thus, terms such as $f(f)$ are ruled out, even if f is the identity function.

- *Polymorphically typed lambda calculus.* This is an intermediate situation, where we may specify, for instance, that a term has a type of the form $X \to X$ for all X, without actually specifying X.

As we will see, each of these alternatives has dramatically different properties from the others.

1.4 Lambda calculus and computability

In the 1930s, several people were interested in the question: what does it mean for a function $f : \mathbb{N} \to \mathbb{N}$ to be *computable*? An informal definition of computability is that there should be a pencil-and-paper method allowing a trained person to calculate $f(n)$, for any given n. The concept of a pencil-and-paper method is not so easy to formalize. Three different researchers attempted to do so, resulting in the following definitions of computability:

1. **Turing** defined an idealized computer we now call a *Turing machine*, and postulated that a function is computable (in the intuitive sense) if and only if it can be computed by such a machine.

2. **Gödel** defined the class of *general recursive functions* as the smallest set of functions containing all the constant functions, the successor function, and closed under certain operations (such as compositions and recursion). He postulated that a function is computable (in the intuitive sense) if and only if it is general recursive.

3. **Church** defined an idealized programming language called the *lambda calculus*, and postulated that a function is computable (in the intuitive sense) if and only if it can be written as a lambda term.

It was proved by Church, Kleene, Rosser, and Turing that all three computational models were equivalent to each other, i.e., each model defines the same class of computable functions. Whether or not they are equivalent to the "intuitive" notion of computability is a question that cannot be answered, because there is no formal definition of "intuitive computability". The assertion that they are in fact equivalent to intuitive computability is known as the *Church-Turing thesis*.

1.5 Connections to computer science

The lambda calculus is a very idealized programming language; arguably, it is the simplest possible programming language that is Turing complete. Because of its simplicity, it is a useful tool for defining and proving properties of programs.

Many real-world programming languages can be regarded as extensions of the lambda calculus. This is true for all *functional programming languages*, a class that includes Lisp, Scheme, Haskell, and ML. Such languages combine the lambda calculus with additional features, such as data types, input/output, side effects, updatable memory, object oriented features, etc. The lambda calculus provides a vehicle for studying such extensions, in isolation and jointly, to see how they will affect each other, and to prove properties of programming languages (such as: a well-formed program will not crash).

The lambda calculus is also a tool used in compiler construction, see e.g. [8, 9].

1.6 Connections to logic

In the 19th and early 20th centuries, there was a philosophical dispute among mathematicians about what a proof is. The so-called *constructivists*, such as Brouwer and Heyting, believed that to prove that a mathematical object exists, one must be able to construct it explicitly. *Classical logicians*, such as Hilbert, held that it is sufficient to derive a contradiction from the assumption that it doesn't exist.

Ironically, one of the better-known examples of a proof that isn't constructive is Brouwer's proof of his own fixed point theorem, which states that every continuous function on the unit disk has a fixed point. The proof is by contradiction and does not give any information on the location of the fixed point.

The connection between lambda calculus and constructive logics is via the "proofs-as-programs" paradigm. To a constructivist, a proof (of an existence statement) must be a "construction", i.e., a program. The lambda calculus is a notation for such programs, and it can also be used as a notation for (constructive) proofs.

For the most part, constructivism has not prevailed as a philosophy in mainstream mathematics. However, there has been renewed interest in constructivism in the second half of the 20th century. The reason is that constructive proofs give more information than classical ones, and in particular, they allow one to compute solutions to problems (as opposed to merely knowing the existence of a solution). The resulting algorithms can be useful in computational mathematics, for instance in computer algebra systems.

1.7 Connections to mathematics

One way to study the lambda calculus is to give mathematical models of it, i.e., to provide spaces in which lambda terms can be given meaning. Such models are constructed using methods from algebra, partially ordered sets, topology, category theory, and other areas of mathematics.

2 The untyped lambda calculus

2.1 Syntax

The lambda calculus is a *formal language*. The expressions of the language are called *lambda terms*, and we will give rules for manipulating them.

Definition. Assume given an infinite set $\mathcal{V}$ of *variables*, denoted by x, y, z etc. The set of lambda terms is given by the following Backus-Naur Form:

$$\text{Lambda terms:}\quad M, N ::= x \mid (MN) \mid (\lambda x.M)$$

The above Backus-Naur Form (BNF) is a convenient abbreviation for the following equivalent, more traditionally mathematical definition:

Definition. Assume given an infinite set $\mathcal{V}$ of variables. Let A be an alphabet consisting of the elements of $\mathcal{V}$, and the special symbols "(", ")", "λ", and ".". Let A^* be the set of strings (finite sequences) over the alphabet A. The set of lambda terms is the smallest subset $\Lambda \subseteq A^*$ such that:

- Whenever $x \in \mathcal{V}$ then $x \in \Lambda$.
- Whenever $M, N \in \Lambda$ then $(MN) \in \Lambda$.
- Whenever $x \in \mathcal{V}$ and $M \in \Lambda$ then $(\lambda x.M) \in \Lambda$.

Comparing the two equivalent definitions, we see that the Backus-Naur Form is a convenient notation because: (1) the definition of the alphabet can be left implicit, (2) the use of distinct meta-symbols for different syntactic classes (x, y, z for variables and M, N for terms) eliminates the need to explicitly quantify over the sets $\mathcal{V}$ and Λ. In the future, we will always present syntactic definitions in the BNF style.

The following are some examples of lambda terms:

$$(\lambda x.x) \qquad ((\lambda x.(xx))(\lambda y.(yy))) \qquad (\lambda f.(\lambda x.(f(fx))))$$

Note that in the definition of lambda terms, we have built in enough mandatory parentheses to ensure that every term $M \in \Lambda$ can be uniquely decomposed into subterms. This means, each term $M \in \Lambda$ is of precisely one of the forms x, (MN), $(\lambda x.M)$. Terms of these three forms are called *variables*, *applications*, and *lambda abstractions*, respectively.

We use the notation (MN), rather than $M(N)$, to denote the application of a function M to an argument N. Thus, in the lambda calculus, we write (fx) instead of the more traditional $f(x)$. This allows us to economize more efficiently on the use of parentheses. To avoid having to write an excessive number of parentheses, we establish the following conventions for writing lambda terms:

Convention.

- We omit outermost parentheses. For instance, we write MN instead of (MN).

- Applications associate to the left; thus, MNP means $(MN)P$. This is convenient when applying a function to a number of arguments, as in $fxyz$, which means $((fx)y)z$. Applying a function to multiple arguments by applying it to one argument at a time is also known as "currying" the function.

- The body of a lambda abstraction (the part after the dot) extends as far to the right as possible. In particular, $\lambda x.MN$ means $\lambda x.(MN)$, and not $(\lambda x.M)N$.

- Multiple lambda abstractions can be contracted; thus $\lambda xyz.M$ will abbreviate $\lambda x.\lambda y.\lambda z.M$.

It is important to note that this convention is only for notational convenience; it does not affect the "official" definition of lambda terms.

Exercise 3. (a) Write the following terms with as few parenthesis as possible, without changing the meaning or structure of the terms:

(i) $(\lambda x.(\lambda y.(\lambda z.((xz)(yz)))))$,

(ii) $(((ab)(cd))((ef)(gh)))$,

(iii) $(\lambda x.((\lambda y.(yx))(\lambda v.v)z)u)(\lambda w.w)$.

(b) Restore all the dropped parentheses in the following terms, without changing the meaning or structure of the terms:

(i) $xxxx$,

(ii) $\lambda x.x\lambda y.y$,

(iii) $\lambda x.(x\lambda y.yxx)x$.

2.2 Free and bound variables, α-equivalence

In our informal discussion of lambda terms, we have already pointed out that the terms $\lambda x.x$ and $\lambda y.y$, which differ only in the name of their bound variable, are essentially the same. We will say that such terms are α-equivalent, and we write $M =_\alpha N$. In the rare event that we want to say that two terms are precisely equal, symbol for symbol, we say that M and N are *identical* and we write $M \equiv N$. We reserve "$=$" as a generic symbol used for different purposes.

An occurrence of a variable x inside a term of the form $\lambda x.N$ is said to be *bound*. The corresponding λx is called a *binder*, and we say that the subterm N is the *scope* of the binder. A variable occurrence that is not bound is *free*. Thus, for example, in the term

$$M \equiv (\lambda x.xy)(\lambda y.yz),$$

x is bound, but z is free. The variable y has both a free and a bound occurrence. The set of free variables of M is $\{y, z\}$.

More generally, the set of free variables of a term M is denoted $FV(M)$, and it is defined formally as follows:

$$\begin{array}{lcl} FV(x) & = & \{x\}, \\ FV(MN) & = & FV(M) \cup FV(N), \\ FV(\lambda x.M) & = & FV(M) \setminus \{x\}. \end{array}$$

This definition is an example of a definition by recursion on terms. In other words, in defining $FV(M)$, we assume that we have already defined $FV(N)$ for all subterms of M. We will often encounter such recursive definitions, as well as inductive proofs.

Before we can formally define α-equivalence, we need to define what it means to *rename* a variable in a term. If x, y are variables, and M is a term, we write $M\{y/x\}$ for the result of renaming x as y in M. Renaming is formally defined as follows:

$$\begin{array}{lcll} x\{y/x\} & \equiv & y, & \\ z\{y/x\} & \equiv & z, & \text{if } x \neq z, \\ (MN)\{y/x\} & \equiv & (M\{y/x\})(N\{y/x\}), & \\ (\lambda x.M)\{y/x\} & \equiv & \lambda y.(M\{y/x\}), & \\ (\lambda z.M)\{y/x\} & \equiv & \lambda z.(M\{y/x\}), & \text{if } x \neq z. \end{array}$$

Note that this kind of renaming replaces all occurrences of x by y, whether free, bound, or binding. We will only apply it in cases where y does not already occur in M.

Finally, we are in a position to formally define what it means for two terms to be "the same up to renaming of bound variables":

Definition. We define *α-equivalence* to be the smallest congruence relation $=_\alpha$ on lambda terms, such that for all terms M and all variables y that do not occur in M,

$$\lambda x.M =_\alpha \lambda y.(M\{y/x\}).$$

Recall that a relation on lambda terms is an equivalence relation if it satisfies rules (*refl*), (*symm*), and (*trans*). It is a congruence if it also satisfies rules (*cong*) and (ξ). Thus, by definition, α-equivalence is the smallest relation on lambda terms satisfying the six rules in Table 1.

It is easy to prove by induction that any lambda term is α-equivalent to another term in which the names of all bound variables are distinct from each other and from any free variables. Thus, when we manipulate lambda terms in theory and in practice, we can (and will) always assume without loss of generality that bound variables have been renamed to be distinct. This convention is called *Barendregt's variable convention*.

As a remark, the notions of free and bound variables and α-equivalence are of course not particular to the lambda calculus; they appear in many standard

(*refl*)	$\dfrac{}{M = M}$	(*cong*)	$\dfrac{M = M' \qquad N = N'}{MN = M'N'}$
(*symm*)	$\dfrac{M = N}{N = M}$	(ξ)	$\dfrac{M = M'}{\lambda x.M = \lambda x.M'}$
(*trans*)	$\dfrac{M = N \qquad N = P}{M = P}$	(α)	$\dfrac{y \notin M}{\lambda x.M = \lambda y.(M\{y/x\})}$

Table 1: The rules for alpha-equivalence

mathematical notations, as well as in computer science. Here are four examples where the variable x is bound.

$$\int_0^1 x^2\, dx$$

$$\sum_{x=1}^{10} \frac{1}{x}$$

$$\lim_{x\to\infty} e^{-x}$$

```
int succ(int x) { return x+1; }
```

2.3 Substitution

In the previous section, we defined a renaming operation, which allowed us to replace a variable by another variable in a lambda term. Now we turn to a less trivial operation, called *substitution*, which allows us to replace a variable by a lambda term. We will write $M[N/x]$ for the result of replacing x by N in M. The definition of substitution is complicated by two circumstances:

1. We should only replace *free* variables. This is because the names of bound variables are considered immaterial, and should not affect the result of a substitution. Thus, $x(\lambda xy.x)[N/x]$ is $N(\lambda xy.x)$, and not $N(\lambda xy.N)$.

2. We need to avoid unintended "capture" of free variables. Consider for example the term $M \equiv \lambda x.yx$, and let $N \equiv \lambda z.xz$. Note that x is free in N and bound in M. What should be the result of substituting N for y in M? If we do this naively, we get

$$M[N/y] = (\lambda x.yx)[N/y] = \lambda x.Nx = \lambda x.(\lambda z.xz)x.$$

However, this is not what we intended, since the variable x was free in N, and during the substitution, it got bound. We need to account for the fact that the x that was bound in M was not the "same" x as the one that was free in N. The proper thing to do is to rename the bound variable *before* the substitution:

$$M[N/y] = (\lambda x'.yx')[N/y] = \lambda x'.Nx' = \lambda x'.(\lambda z.xz)x'.$$

Thus, the operation of substitution forces us to sometimes rename a bound variable. In this case, it is best to pick a variable from $\mathcal{V}$ that has not been used yet as the new name of the bound variable. A variable that is currently unused is called *fresh*. The reason we stipulated that the set $\mathcal{V}$ is infinite was to make sure a fresh variable is always available when we need one.

Definition. The (capture-avoiding) *substitution* of N for free occurrences of x in M, in symbols $M[N/x]$, is defined as follows:

$$
\begin{array}{lll}
x[N/x] & \equiv N, & \\
y[N/x] & \equiv y, & \text{if } x \neq y, \\
(MP)[N/x] & \equiv (M[N/x])(P[N/x]), & \\
(\lambda x.M)[N/x] & \equiv \lambda x.M, & \\
(\lambda y.M)[N/x] & \equiv \lambda y.(M[N/x]), & \text{if } x \neq y \text{ and } y \notin FV(N), \\
(\lambda y.M)[N/x] & \equiv \lambda y'.(M\{y'/y\}[N/x]), & \text{if } x \neq y,\, y \in FV(N), \text{ and } y' \text{ fresh.}
\end{array}
$$

This definition has one technical flaw: in the last clause, we did not specify which fresh variable to pick, and thus, technically, substitution is not well-defined. One way to solve this problem is to declare all lambda terms to be identified up to α-equivalence, and to prove that substitution is in fact well-defined modulo α-equivalence. Another way would be to specify which variable y' to choose: for instance, assume that there is a well-ordering on the set $\mathcal{V}$ of variables, and stipulate that y' should be chosen to be the least variable that does not occur in either M or N.

2.4 Introduction to β-reduction

Convention. From now on, unless stated otherwise, we identify lambda terms up to α-equivalence. This means, when we speak of lambda terms being "equal", we mean that they are α-equivalent. Formally, we regard lambda terms as equivalence classes modulo α-equivalence. We will often use the ordinary equality symbol $M = N$ to denote α-equivalence.

The process of evaluating lambda terms by "plugging arguments into functions" is called *β-reduction*. A term of the form $(\lambda x.M)N$, which consists of a lambda abstraction applied to another term, is called a *β-redex*. We say that it *reduces* to $M[N/x]$, and we call the latter term the *reduct*. We reduce lambda terms by finding a subterm that is a redex, and then replacing that redex by its reduct. We repeat this as many times as we like, or until there are no more redexes left to reduce. A lambda term without any β-redexes is said to be in *β-normal form*.

For example, the lambda term $(\lambda x.y)((\lambda z.zz)(\lambda w.w))$ can be reduced as follows. Here, we underline each redex just before reducing it:

$$
\begin{array}{rll}
(\lambda x.y)(\underline{(\lambda z.zz)(\lambda w.w)}) & \to_\beta & (\lambda x.y)(\underline{(\lambda w.w)(\lambda w.w)}) \\
& \to_\beta & \underline{(\lambda x.y)(\lambda w.w)} \\
& \to_\beta & y.
\end{array}
$$

The last term, y, has no redexes and is thus in normal form. We could reduce the same term differently, by choosing the redexes in a different order:

$$\underline{(\lambda x.y)((\lambda z.zz)(\lambda w.w))} \quad \to_\beta \quad y.$$

As we can see from this example:

- reducing a redex can create new redexes,
- reducing a redex can delete some other redexes,
- the number of steps that it takes to reach a normal form can vary, depending on the order in which the redexes are reduced.

We can also see that the final result, y, does not seem to depend on the order in which the redexes are reduced. In fact, this is true in general, as we will prove later.

If M and M' are terms such that $M \twoheadrightarrow_\beta M'$, and if M' is in normal form, then we say that M *evaluates* to M'.

Not every term evaluates to something; some terms can be reduced forever without reaching a normal form. The following is an example:

$$\begin{array}{rcl} (\lambda x.xx)(\lambda y.yyy) & \to_\beta & (\lambda y.yyy)(\lambda y.yyy) \\ & \to_\beta & (\lambda y.yyy)(\lambda y.yyy)(\lambda y.yyy) \\ & \to_\beta & \ldots \end{array}$$

This example also shows that the size of a lambda term need not decrease during reduction; it can increase, or remain the same. The term $(\lambda x.xx)(\lambda x.xx)$, which we encountered in Section 1, is another example of a lambda term that does not reach a normal form.

2.5 Formal definitions of β-reduction and β-equivalence

The concept of β-reduction can be defined formally as follows:

Definition. We define *single-step β-reduction* to be the smallest relation $\to_\beta$ on terms satisfying:

$$\begin{array}{ll} (\beta) & \dfrac{}{(\lambda x.M)N \to_\beta M[N/x]} \\[2ex] (\mathit{cong}_1) & \dfrac{M \to_\beta M'}{MN \to_\beta M'N} \\[2ex] (\mathit{cong}_2) & \dfrac{N \to_\beta N'}{MN \to_\beta MN'} \\[2ex] (\xi) & \dfrac{M \to_\beta M'}{\lambda x.M \to_\beta \lambda x.M'} \end{array}$$

Thus, $M \to_\beta M'$ iff M' is obtained from M by reducing a *single* β-redex of M.

Definition. We write $M \twoheadrightarrow_\beta M'$ if M reduces to M' in zero or more steps. Formally, $\twoheadrightarrow_\beta$ is defined to be the reflexive transitive closure of $\to_\beta$, i.e., the smallest reflexive transitive relation containing $\to_\beta$.

Finally, β-equivalence is obtained by allowing reduction steps as well as inverse reduction steps, i.e., by making $\to_\beta$ symmetric:

Definition. We write $M =_\beta M'$ if M can be transformed into M' by zero or more reduction steps and/or inverse reduction steps. Formally, $=_\beta$ is defined to be the reflexive symmetric transitive closure of $\to_\beta$, i.e., the smallest equivalence relation containing $\to_\beta$.

Exercise 4. A slightly different way to define β-equivalence is as the smallest equivalence relation $=_\beta$ on terms satisfying:

$$(\beta)\qquad \frac{}{(\lambda x.M)N =_\beta M[N/x]}$$

$$(cong_1)\qquad \frac{M =_\beta M'}{MN =_\beta M'N}$$

$$(cong_2)\qquad \frac{N =_\beta N'}{MN =_\beta MN'}$$

$$(\xi)\qquad \frac{M =_\beta M'}{\lambda x.M =_\beta \lambda x.M'}$$

Prove that the two definitions are equivalent.

3 Programming in the untyped lambda calculus

One of the amazing facts about the untyped lambda calculus is that we can use it to encode data, such as booleans and natural numbers, as well as programs that operate on the data. This can be done purely within the lambda calculus, without adding any additional syntax or axioms.

We will often have occasion to give names to particular lambda terms; we will usually use boldface letters for such names.

3.1 Booleans

We begin by defining two lambda terms to encode the truth values "true" and "false":

$$\begin{array}{rcl} \mathbf{T} & = & \lambda xy.x \\ \mathbf{F} & = & \lambda xy.y \end{array}$$

Let **and** be the term $\lambda ab.ab\mathbf{F}$. Verify the following:

$$\begin{array}{lcl} \textbf{and TT} & \twoheadrightarrow_\beta & \mathbf{T} \\ \textbf{and TF} & \twoheadrightarrow_\beta & \mathbf{F} \\ \textbf{and FT} & \twoheadrightarrow_\beta & \mathbf{F} \\ \textbf{and FF} & \twoheadrightarrow_\beta & \mathbf{F} \end{array}$$

Note that $\mathbf{T}$ and $\mathbf{F}$ are normal forms, so we can really say that a term such as **and TT** *evaluates* to $\mathbf{T}$. We say that **and** *encodes* the boolean function "and". It is understood that this coding is with respect to the particular coding of "true" and "false". We don't claim that **and** MN evaluates to anything meaningful if M or N are terms other than $\mathbf{T}$ and $\mathbf{F}$.

Incidentally, there is nothing unique about the term $\lambda ab.ab\mathbf{F}$. It is one of many possible ways of encoding the "and" function. Another possibility is $\lambda ab.bab$.

Exercise 5. Find lambda terms **or** and **not** that encode the boolean functions "or" and "not". Can you find more than one term?

Moreover, we define the term **if_then_else** $= \lambda x.x$. This term behaves like an "if-then-else" function — specifically, we have

$$\begin{array}{lcl} \textbf{if\_then\_else}\ \mathbf{T}MN & \twoheadrightarrow_\beta & M \\ \textbf{if\_then\_else}\ \mathbf{F}MN & \twoheadrightarrow_\beta & N \end{array}$$

for all lambda terms M, N.

3.2 Natural numbers

If f and x are lambda terms, and $n \geqslant 0$ a natural number, write $f^n x$ for the term $f(f(\ldots(fx)\ldots))$, where f occurs n times. For each natural number n, we define a lambda term $\overline{n}$, called the *nth Church numeral*, as $\overline{n} = \lambda fx.f^n x$. Here are the first few Church numerals:

$$\begin{array}{lcl} \overline{0} & = & \lambda fx.x \\ \overline{1} & = & \lambda fx.fx \\ \overline{2} & = & \lambda fx.f(fx) \\ \overline{3} & = & \lambda fx.f(f(fx)) \\ \ldots & & \end{array}$$

This particular way of encoding the natural numbers is due to Alonzo Church, who was also the inventor of the lambda calculus. Note that $\overline{0}$ is in fact the same term as $\mathbf{F}$; thus, when interpreting a lambda term, we should know ahead of time whether to interpret the result as a boolean or a numeral.

The successor function can be defined as follows: **succ** $= \lambda nfx.f(nfx)$.

What does this term compute when applied to a numeral?

$$\begin{array}{rcl}
\mathbf{succ}\ \overline{n} & = & (\lambda nfx.f(nfx))(\lambda fx.f^n x) \\
& \to_\beta & \lambda fx.f((\lambda fx.f^n x)fx) \\
& \twoheadrightarrow_\beta & \lambda fx.f(f^n x) \\
& = & \lambda fx.f^{n+1}x \\
& = & \overline{n+1}
\end{array}$$

Thus, we have proved that the term **succ** does indeed encode the successor function, when applied to a numeral. Here are possible definitions of addition and multiplication:

$$\begin{array}{rcl}
\mathbf{add} & = & \lambda nmfx.nf(mfx) \\
\mathbf{mult} & = & \lambda nmf.n(mf).
\end{array}$$

Exercise 6. (a) Manually evaluate the lambda terms **add** $\overline{2}\,\overline{3}$ and **mult** $\overline{2}\,\overline{3}$.

(b) Prove that **add** $\overline{n}\,\overline{m} \twoheadrightarrow_\beta \overline{n+m}$, for all natural numbers n, m.

(c) Prove that **mult** $\overline{n}\,\overline{m} \twoheadrightarrow_\beta \overline{n \cdot m}$, for all natural numbers n, m.

Definition. Suppose $f : \mathbb{N}^k \to \mathbb{N}$ is a k-ary function on the natural numbers, and that M is a lambda term. We say that M *(numeralwise) represents* f if for all $n_1, \ldots, n_k \in \mathbb{N}$,

$$M\,\overline{n_1} \ldots \overline{n_k} \twoheadrightarrow_\beta \overline{f(n_1, \ldots, n_k)}\,.$$

This definition makes explicit what it means to be an "encoding". We can say, for instance, that the term **add** $= \lambda nmfx.nf(mfx)$ represents the addition function. The definition generalizes easily to boolean functions, or functions of other data types.

Often handy is the function **iszero** from natural numbers to booleans, which is defined by

$$\begin{array}{rcl}
\mathbf{iszero}\,(0) & = & \text{true} \\
\mathbf{iszero}\,(n) & = & \text{false}, \quad \text{if } n \neq 0.
\end{array}$$

Convince yourself that the following term is a representation of this function:

$$\mathbf{iszero} = \lambda nxy.n(\lambda z.y)x.$$

Exercise 7. Find lambda terms that represent each of the following functions:

(a) $f(n) = (n+3)^2$,

(b) $f(n) = \begin{cases} \text{true} & \text{if } n \text{ is even,} \\ \text{false} & \text{if } n \text{ is odd,} \end{cases}$

(c) $\mathbf{exp}\,(n, m) = n^m$,

(d) $\mathbf{pred}\,(n) = n - 1$.

Note: part (d) is not easy. In fact, Church believed for a while that it was impossible, until his student Kleene found a solution. (In fact, Kleene said he found the solution while having his wisdom teeth pulled, so his trick for defining the predecessor function is sometimes referred to as the "wisdom teeth trick".)

We have seen how to encode some simple boolean and arithmetic functions. However, we do not yet have a systematic method of constructing such functions. What we need is a mechanism for defining more complicated functions from simple ones. Consider for example the factorial function, defined by:

$$\begin{array}{rcl} 0! & = & 1 \\ n! & = & n \cdot (n-1)!, \quad \text{if } n \neq 0. \end{array}$$

The encoding of such functions in the lambda calculus is the subject of the next section. It is related to the concept of a fixed point.

3.3 Fixed points and recursive functions

Suppose f is a function. We say that x is a *fixed point* of f if $f(x) = x$. In arithmetic and calculus, some functions have fixed points, while others don't. For instance, $f(x) = x^2$ has two fixed points 0 and 1, whereas $f(x) = x + 1$ has no fixed points. Some functions have infinitely many fixed points, notably $f(x) = x$.

We apply the notion of fixed points to the lambda calculus. If F and N are lambda terms, we say that N is a fixed point of F if $FN =_\beta N$. The lambda calculus contrasts with arithmetic in that *every* lambda term has a fixed point. This is perhaps the first surprising fact about the lambda calculus we learn in this course.

Theorem 3.1. *In the untyped lambda calculus, every term F has a fixed point.*

Proof. Let $A = \lambda xy.y(xxy)$, and define $\Theta = AA$. Now suppose F is any lambda term, and let $N = \Theta F$. We claim that N is a fixed point of F. This is shown by the following calculation:

$$\begin{array}{rcl} N & = & \Theta F \\ & = & AAF \\ & = & (\lambda xy.y(xxy))AF \\ & \twoheadrightarrow_\beta & F(AAF) \\ & = & F(\Theta F) \\ & = & FN. \end{array}$$

□

The term Θ used in the proof is called *Turing's fixed point combinator*.

The importance of fixed points lies in the fact that they allow us to solve *equations*. After all, finding a fixed point for f is the same thing as solving the equation $x = f(x)$. This covers equations with an arbitrary right-hand side, whose left-hand side is x. From the above theorem, we know that we can always solve such equations in the lambda calculus.

To see how to apply this idea, consider the question from the last section, namely, how to define the factorial function. The most natural definition of the factorial function is recursive, and we can write it in the lambda calculus as follows:

$$\textbf{fact}\,n \quad = \quad \textbf{if\_then\_else}\,(\textbf{iszero}\,n)(\overline{1})(\textbf{mult}\,n(\textbf{fact}\,(\textbf{pred}\,n)))$$

Here we have used various abbreviations for lambda terms that were introduced in the previous section. The evident problem with a recursive definition such as this one is that the term to be defined, **fact**, appears both on the left- and the right-hand side. In other words, to find **fact** requires solving an equation!

We now apply our newfound knowledge of how to solve fixed point equations in the lambda calculus. We start by rewriting the problem slightly:

$$\begin{array}{rcl}\textbf{fact} & = & \lambda n.\,\textbf{if\_then\_else}\,(\textbf{iszero}\,n)(\overline{1})(\textbf{mult}\,n(\textbf{fact}\,(\textbf{pred}\,n)))\\ \textbf{fact} & = & (\lambda f.\lambda n.\,\textbf{if\_then\_else}\,(\textbf{iszero}\,n)(\overline{1})(\textbf{mult}\,n(f(\textbf{pred}\,n))))\,\textbf{fact}\end{array}$$

Let us temporarily write F for the term

$$\lambda f.\lambda n.\,\textbf{if\_then\_else}\,(\textbf{iszero}\,n)(\overline{1})(\textbf{mult}\,n(f(\textbf{pred}\,n))).$$

Then the last equation becomes $\textbf{fact} = F\,\textbf{fact}$, which is a fixed point equation. We can solve it up to β-equivalence, by letting

$$\begin{array}{rcl}\textbf{fact} & = & \boldsymbol{\Theta} F\\ & = & \boldsymbol{\Theta}(\lambda f.\lambda n.\,\textbf{if\_then\_else}\,(\textbf{iszero}\,n)(\overline{1})(\textbf{mult}\,n(f(\textbf{pred}\,n))))\end{array}$$

Note that **fact** has disappeared from the right-hand side. The right-hand side is a closed lambda term that represents the factorial function. (A lambda term is called *closed* if it contains no free variables).

To see how this definition works in practice, let us evaluate $\textbf{fact}\,\overline{2}$. Recall from the proof of Theorem 3.1 that $\boldsymbol{\Theta} F \twoheadrightarrow_\beta F(\boldsymbol{\Theta} F)$, therefore $\textbf{fact} \twoheadrightarrow_\beta F\,\textbf{fact}$.

$$\begin{array}{ll}\textbf{fact}\,\overline{2} \twoheadrightarrow_\beta F\,\textbf{fact}\,\overline{2} & \\ \quad \twoheadrightarrow_\beta \textbf{if\_then\_else}\,(\textbf{iszero}\,\overline{2})(\overline{1})(\textbf{mult}\,\overline{2}(\textbf{fact}\,(\textbf{pred}\,\overline{2}))) & \\ \quad \twoheadrightarrow_\beta \textbf{if\_then\_else}\,(\textbf{F})(\overline{1})(\textbf{mult}\,\overline{2}(\textbf{fact}\,(\textbf{pred}\,\overline{2}))) & \\ \quad \twoheadrightarrow_\beta \textbf{mult}\,\overline{2}(\textbf{fact}\,(\textbf{pred}\,\overline{2})) & \\ \quad \twoheadrightarrow_\beta \textbf{mult}\,\overline{2}(\textbf{fact}\,\overline{1}) & \\ \quad \twoheadrightarrow_\beta \textbf{mult}\,\overline{2}(F\,\textbf{fact}\,\overline{1}) & \\ \quad \twoheadrightarrow_\beta \ldots & \\ \quad \twoheadrightarrow_\beta \textbf{mult}\,\overline{2}(\textbf{mult}\,\overline{1}(\textbf{fact}\,\overline{0})) & \\ \quad \twoheadrightarrow_\beta \textbf{mult}\,\overline{2}(\textbf{mult}\,\overline{1}(F\,\textbf{fact}\,\overline{0})) & \\ \quad \twoheadrightarrow_\beta \textbf{mult}\,\overline{2}(\textbf{mult}\,\overline{1}(\textbf{if\_then\_else}\,(\textbf{iszero}\,\overline{0})(\overline{1})(\textbf{mult}\,\overline{0}(\textbf{fact}\,(\textbf{pred}\,\overline{0}))))) & \\ \quad \twoheadrightarrow_\beta \textbf{mult}\,\overline{2}(\textbf{mult}\,\overline{1}(\textbf{if\_then\_else}\,(\textbf{T})(\overline{1})(\textbf{mult}\,\overline{0}(\textbf{fact}\,(\textbf{pred}\,\overline{0}))))) & \end{array}$$

$$\begin{array}{ll} \twoheadrightarrow_\beta & \textbf{mult}\ \overline{2}(\textbf{mult}\ \overline{1}\,\overline{1}) \\ \twoheadrightarrow_\beta & \overline{2} \end{array}$$

Note that this calculation, while messy, is completely mechanical. You can easily convince yourself that **fact** $\overline{3}$ reduces to **mult** $\overline{3}$(**fact** $\overline{2}$), and therefore, by the above calculation, to **mult** $\overline{3}\,\overline{2}$, and finally to $\overline{6}$. It is now a matter of a simple induction to prove that **fact** $\overline{n} \twoheadrightarrow_\beta \overline{n!}$, for any n.

Exercise 8. Write a lambda term that represents the Fibonacci function, defined by

$$f(0) = 1, \qquad f(1) = 1, \qquad f(n+2) = f(n+1) + f(n), \text{for } n \geqslant 2$$

Exercise 9. Write a lambda term that represents the characteristic function of the prime numbers, i.e., $f(n) =$ true if n is prime, and false otherwise.

Exercise 10. We have remarked at the beginning of this section that the number-theoretic function $f(x) = x + 1$ does not have a fixed point. On the other hand, the lambda term $F = \lambda x.\,\textbf{succ}\ x$, which represents the same function, does have a fixed point by Theorem 3.1. How can you reconcile the two statements?

Exercise 11. The first fixed point combinator for the lambda calculus was discovered by Curry. Curry's fixed point combinator, which is also called the *paradoxical fixed point combinator*, is the term $\mathbf{Y} = \lambda f.(\lambda x.f(xx))(\lambda x.f(xx))$.

(a) Prove that this is indeed a fixed point combinator, i.e., that $\mathbf{Y}F$ is a fixed point of F, for any term F.

(b) Turing's fixed point combinator not only satisfies $\mathbf{\Theta}F =_\beta F(\mathbf{\Theta}F)$, but also $\mathbf{\Theta}F \twoheadrightarrow_\beta F(\mathbf{\Theta}F)$. We used this fact in evaluating **fact** $\overline{2}$. Does an analogous property hold for $\mathbf{Y}$? Does this affect the outcome of the evaluation of **fact** $\overline{2}$?

(c) Can you find another fixed point combinator, besides Curry's and Turing's?

3.4 Other data types: pairs, tuples, lists, trees, etc.

So far, we have discussed lambda terms that represented functions on booleans and natural numbers. However, it is easily possible to encode more general data structures in the untyped lambda calculus. Pairs and tuples are of interest to everybody. The examples of lists and trees are primarily interesting to people with experience in a list-processing language such as LISP or PROLOG; you can safely ignore these examples if you want to.

Pairs. If M and N are lambda terms, we define the pair $\langle M, N\rangle$ to be the lambda term $\lambda z.zMN$. We also define two terms $\pi_1 = \lambda p.p(\lambda xy.x)$ and $\pi_2 = \lambda p.p(\lambda xy.y)$. We observe the following:

$$\begin{array}{lll} \pi_1\langle M, N\rangle & \twoheadrightarrow_\beta & M \\ \pi_2\langle M, N\rangle & \twoheadrightarrow_\beta & N \end{array}$$

The terms π_1 and π_2 are called the left and right *projections*.

Tuples. The encoding of pairs can easily be extended to arbitrary n-tuples. If $M_1, \ldots, M_n$ are terms, we define the n-tuple $\langle M_1, \ldots, M_n \rangle$ as the lambda term $\lambda z.zM_1 \ldots M_n$, and we define the ith projection $\pi_i^n = \lambda p.p(\lambda x_1 \ldots x_n.x_i)$. Then

$$\pi_i^n \langle M_1, \ldots, M_n \rangle \quad \twoheadrightarrow_\beta \quad M_i, \text{for all } 1 \leqslant i \leqslant n.$$

Lists. A list is different from a tuple, because its length is not necessarily fixed. A list is either empty ("nil"), or else it consists of a first element (the "head") followed by another list (the "tail"). We write **nil** for the empty list, and $H :: T$ for the list whose head is H and whose tail is T. So, for instance, the list of the first three numbers can be written as $1 :: (2 :: (3 :: \textbf{nil}\,))$. We usually omit the parentheses, where it is understood that "::" associates to the right. Note that every list ends in **nil**.

In the lambda calculus, we can define $\textbf{nil} = \lambda xy.y$ and $H :: T = \lambda xy.xHT$. Here is a lambda term that adds a list of numbers:

$$\textbf{addlist}\ l = l(\lambda h\, t.\,\textbf{add}\ h(\textbf{addlist}\ t))(\overline{0}).$$

Of course, this is a recursive definition, and must be translated into an actual lambda term by the method of Section 3.3. In the definition of **addlist**, l and t are lists of numbers, and h is a number. If you are very diligent, you can calculate the sum of last weekend's Canadian lottery results by evaluating the term

$$\textbf{addlist}\,(\overline{4} :: \overline{22} :: \overline{24} :: \overline{32} :: \overline{42} :: \overline{43} :: \textbf{nil}\,).$$

Note that lists enable us to give an alternative encoding of the natural numbers: We can encode a natural number as a list of booleans, which we interpret as the binary digits 0 and 1. Of course, with this encoding, we would have to carefully redesign our basic functions, such as successor, addition, and multiplication. However, if done properly, such an encoding would be a lot more efficient (in terms of number of β-reductions to be performed) than the encoding by Church numerals.

Trees. A binary tree is a data structure that can be one of two things: either a *leaf*, labelled by a natural number, or a *node*, which has a left and a right subtree. We write $\textbf{leaf}\,(N)$ for a leaf labelled N, and $\textbf{node}\,(L, R)$ for a node with left subtree L and right subtree R. We can encode trees as lambda terms, for instance as follows:

$$\textbf{leaf}\,(n) = \lambda xy.xn, \qquad \textbf{node}\,(L, R) = \lambda xy.yLR$$

As an illustration, here is a program (i.e., a lambda term) that adds all the numbers at the leafs of a given tree.

$$\textbf{addtree}\ t = t(\lambda n.n)(\lambda l\, r.\,\textbf{add}\,(\textbf{addtree}\ l)(\textbf{addtree}\ r)).$$

Exercise 12. This is a voluntary programming exercise.

(a) Write a lambda term that calculates the length of a list.

(b) Write a lambda term that calculates the depth (i.e., the nesting level) of a tree. You may need to define a function **max** that calculates the maximum of two numbers.

(c) Write a lambda term that sorts a list of numbers. You may assume given a term **less** that compares two numbers.

4 The Church-Rosser Theorem

4.1 Extensionality, η-equivalence, and η-reduction

In the untyped lambda calculus, any term can be applied to another term. Therefore, any term can be regarded as a function. Consider a term M, not containing the variable x, and consider the term $M' = \lambda x.Mx$. Then for any argument A, we have $MA =_\beta M'A$. So in this sense, M and M' define "the same function". Should M and M' be considered equivalent as terms?

The answer depends on whether we want to accept the principle that "if M and M' define the same function, then M and M' are equal". This is called the principle of *extensionality*, and we have already encountered it in Section 1.1. Formally, the extensionality rule is the following:

$$(\mathit{ext}_\forall)\quad \frac{\forall A.MA = M'A}{M = M'}.$$

In the presence of the axioms (ξ), (*cong*), and (β), it can be easily seen that $MA = M'A$ is true for *all* terms A if and only if $Mx = M'x$, where x is a fresh variable. Therefore, we can replace the extensionality rule by the following equivalent, but simpler rule:

$$(\mathit{ext})\quad \frac{Mx = M'x,\ \text{where}\ x \notin FV(M, M')}{M = M'}.$$

Note that we can apply the extensionality rule in particular to the case where $M' = \lambda x.Mx$, where x is not free in M. As we have remarked above, $Mx =_\beta M'x$, and thus extensionality implies that $M = \lambda x.Mx$. This last equation is called the η-law (eta-law):

$$(\eta)\quad M = \lambda x.Mx,\ \text{where}\ x \notin FV(M).$$

In fact, (η) and (*ext*) are equivalent in the presence of the other axioms of the lambda calculus. We have already seen that (*ext*) and (β) imply (η). Conversely, assume (η), and assume that $Mx = M'x$, for some terms M and M' not containing x freely. Then by (ξ), we have $\lambda x.Mx = \lambda x.M'x$, hence by ($\eta$) and transitivity, $M = M'$. Thus (*ext*) holds.

We note that the η-law does not follow from the axioms and rules of the lambda calculus that we have considered so far. In particular, the terms x and $\lambda y.xy$ are not β-equivalent, although they are clearly η-equivalent. We will prove that $x \neq_\beta \lambda y.xy$ in Corollary 4.5 below.

Single-step η-reduction is the smallest relation $\to_\eta$ satisfying $(cong_1)$, $(cong_2)$, (ξ), and the following axiom (which is the same as the η-law, directed right to left):

$$(\eta) \quad \lambda x.Mx \to_\eta M, \text{ where } x \notin FV(M).$$

Single-step $\beta\eta$-reduction $\to_{\beta\eta}$ is defined as the union of the single-step β- and η-reductions, i.e., $M \to_{\beta\eta} M'$ iff $M \to_\beta M'$ or $M \to_\eta M'$. Multi-step η-reduction $\twoheadrightarrow_\eta$, multi-step $\beta\eta$-reduction $\twoheadrightarrow_{\beta\eta}$, as well as η-equivalence $=_\eta$ and $\beta\eta$-equivalence $=_{\beta\eta}$ are defined in the obvious way as we did for β-reduction and equivalence. We also get the evident notions of η-normal form, $\beta\eta$-normal form, etc.

4.2 Statement of the Church-Rosser Theorem, and some consequences

Theorem (Church and Rosser, 1936). *Let $\twoheadrightarrow$ denote either $\twoheadrightarrow_\beta$ or $\twoheadrightarrow_{\beta\eta}$. Suppose M, N, and P are lambda terms such that $M \twoheadrightarrow N$ and $M \twoheadrightarrow P$. Then there exists a lambda term Z such that $N \twoheadrightarrow Z$ and $P \twoheadrightarrow Z$.*

In pictures, the theorem states that the following diagram can always be completed:

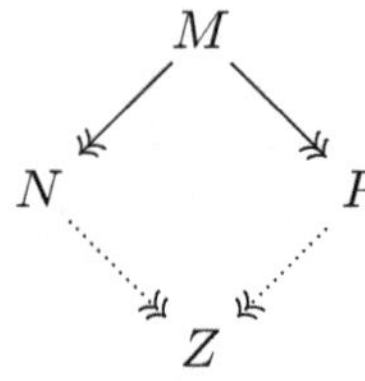

This property is called the *Church-Rosser property*, or *confluence*. Before we prove the Church-Rosser Theorem, let us highlight some of its consequences.

Corollary 4.1. *If $M =_\beta N$ then there exists some Z with $M, N \twoheadrightarrow_\beta Z$. Similarly for $\beta\eta$.*

Proof. Please refer to Figure 1 for an illustration of this proof. Recall that $=_\beta$ is the reflexive symmetric transitive closure of $\to_\beta$. Suppose that $M =_\beta N$. Then there exist $n \geqslant 0$ and terms $M_0, \ldots, M_n$ such that $M = M_0$, $N = M_n$, and for all $i = 1 \ldots n$, either $M_{i-1} \to_\beta M_i$ or $M_i \to_\beta M_{i-1}$. We prove the claim by induction on n. For $n = 0$, we have $M = N$ and there is nothing to show. Suppose the claim has been proven for $n-1$. Then by induction hypothesis, there exists a term Z' such that $M \twoheadrightarrow_\beta Z'$ and $M_{n-1} \twoheadrightarrow_\beta Z'$. Further, we know that either $N \to_\beta M_{n-1}$ or $M_{n-1} \to_\beta N$. In case $N \to_\beta M_{n-1}$, then $N \twoheadrightarrow_\beta Z'$,

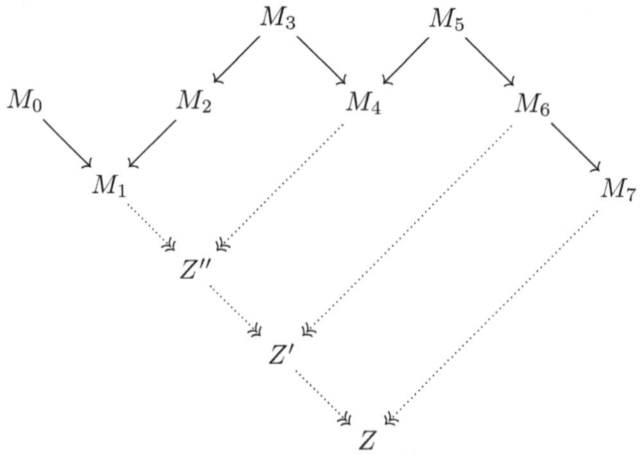

Figure 1: The proof of Corollary 4.1

and we are done. In case $M_{n-1} \to_\beta N$, we apply the Church-Rosser Theorem to M_{n-1}, Z', and N to obtain a term Z such that $Z' \twoheadrightarrow_\beta Z$ and $N \twoheadrightarrow_\beta Z$. Since $M \twoheadrightarrow_\beta Z' \twoheadrightarrow_\beta Z$, we are done. The proof in the case of $\beta\eta$-reduction is identical. □

Corollary 4.2. *If N is a β-normal form and $N =_\beta M$, then $M \twoheadrightarrow_\beta N$, and similarly for $\beta\eta$.*

Proof. By Corollary 4.1, there exists some Z with $M, N \twoheadrightarrow_\beta Z$. But N is a normal form, thus $N =_\alpha Z$. □

Corollary 4.3. *If M and N are β-normal forms such that $M =_\beta N$, then $M =_\alpha N$, and similarly for $\beta\eta$.*

Proof. By Corollary 4.2, we have $M \twoheadrightarrow_\beta N$, but since M is a normal form, we have $M =_\alpha N$. □

Corollary 4.4. *If $M =_\beta N$, then neither or both have a β-normal form. Similarly for $\beta\eta$.*

Proof. Suppose that $M =_\beta N$, and that one of them has a β-normal form. Say, for instance, that M has a normal form Z. Then $N =_\beta Z$, hence $N \twoheadrightarrow_\beta Z$ by Corollary 4.2. □

Corollary 4.5. *The terms x and $\lambda y.xy$ are not β-equivalent. In particular, the η-rule does not follow from the β-rule.*

Proof. The terms x and $\lambda y.xy$ are both β-normal forms, and they are not α-equivalent. It follows by Corollary 4.3 that $x \neq_\beta \lambda y.xy$. □

4.3 Preliminary remarks on the proof of the Church-Rosser Theorem

Consider any binary relation $\to$ on a set, and let $\twoheadrightarrow$ be its reflexive transitive closure. Consider the following three properties of such relations:

(a)
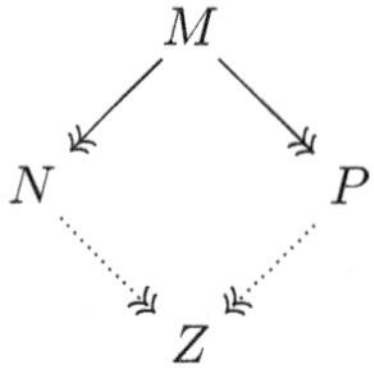

(b)
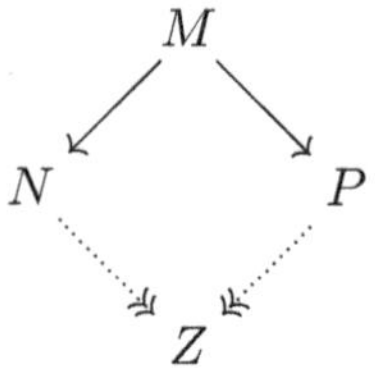

(c)
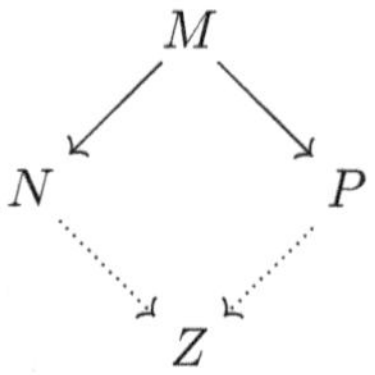

Each of these properties states that for all M, N, P, if the solid arrows exist, then there exists Z such that the dotted arrows exist. The only difference between (a), (b), and (c) is the difference between where $\to$ and $\twoheadrightarrow$ are used.

Property (a) is the Church-Rosser property. Property (c) is called the diamond property (because the diagram is shaped like a diamond).

A naive attempt to prove the Church-Rosser Theorem might proceed as follows: First, prove that the relation $\to_\beta$ satisfies property (b) (this is relatively easy to prove); then use an inductive argument to conclude that it also satisfies property (a).

Unfortunately, this does not work: the reason is that in general, property (b) does not imply property (a)! An example of a relation that satisfies property (b) but not property (a) is shown in Figure 2. In other words, a proof of property (b) is not sufficient in order to prove property (a).

On the other hand, property (c), the diamond property, *does* imply property (a). This is very easy to prove by induction, and the proof is illustrated in Figure 3. But unfortunately, β-reduction does not satisfy property (c), so again we are stuck.

To summarize, we are faced with the following dilemma:

- β-reduction satisfies property (b), but property (b) does not imply property (a).
- Property (c) implies property (a), but β-reduction does not satisfy property (c).

On the other hand, it seems hopeless to prove property (a) directly. In the next section, we will solve this dilemma by defining yet another reduction relation $\triangleright$, with the following properties:

- $\triangleright$ satisfies property (c), and
- the transitive closure of $\triangleright$ is the same as that of $\to_\beta$ (or $\to_{\beta\eta}$).

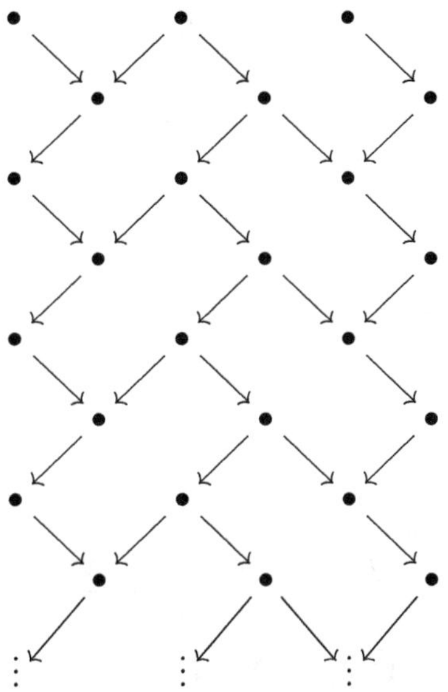

Figure 2: An example of a relation that satisfies property (b), but not property (a)

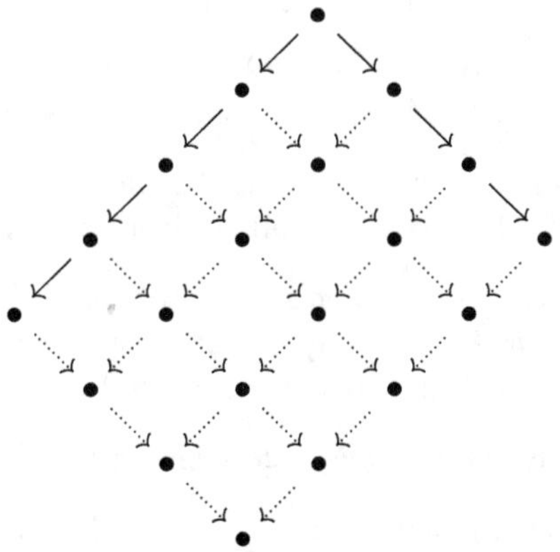

Figure 3: Proof that property (c) implies property (a)

4.4 Proof of the Church-Rosser Theorem

In this section, we will prove the Church-Rosser Theorem for $\beta\eta$-reduction. The proof for β-reduction (without η) is very similar, and in fact slightly simpler, so we omit it here. The proof presented here is due to Tait and Martin-Löf. We begin by defining a new relation $M \rhd M'$ on terms, called *parallel one-step reduction*. We define $\rhd$ to be the smallest relation satisfying

$$
\begin{array}{ll}
(1) & \dfrac{}{x \rhd x} \\[2ex]
(2) & \dfrac{P \rhd P' \qquad N \rhd N'}{PN \rhd P'N'} \\[2ex]
(3) & \dfrac{N \rhd N'}{\lambda x.N \rhd \lambda x.N'} \\[2ex]
(4) & \dfrac{Q \rhd Q' \qquad N \rhd N'}{(\lambda x.Q)N \rhd Q'[N'/x]} \\[2ex]
(5) & \dfrac{P \rhd P', \text{ where } x \not\in FV(P)}{\lambda x.Px \rhd P'}.
\end{array}
$$

Lemma 4.6. *(a) For all M, M', if $M \to_{\beta\eta} M'$ then $M \rhd M'$.*

(b) For all M, M', if $M \rhd M'$ then $M \twoheadrightarrow_{\beta\eta} M'$.

(c) $\twoheadrightarrow_{\beta\eta}$ is the reflexive, transitive closure of $\rhd$.

Proof. (a) First note that we have $P \rhd P$, for any term P. This is easily shown by induction on P. We now prove the claim by induction on a derivation of $M \to_{\beta\eta} M'$. Please refer to pages 11 and 20 for the rules that define $\to_{\beta\eta}$. We make a case distinction based on the last rule used in the derivation of $M \to_{\beta\eta} M'$.

- If the last rule was (β), then $M = (\lambda x.Q)N$ and $M' = Q[N/x]$, for some Q and N. But then $M \rhd M'$ by (4), using the facts $Q \rhd Q$ and $N \rhd N$.

- If the last rule was (η), then $M = \lambda x.Px$ and $M' = P$, for some P such that $x \not\in FV(P)$. Then $M \rhd M'$ follows from (5), using $P \rhd P$.

- If the last rule was $(cong_1)$, then $M = PN$ and $M' = P'N$, for some P, P', and N where $P \to_{\beta\eta} P'$. By induction hypothesis, $P \rhd P'$. From this and $N \rhd N$, it follows immediately that $M \rhd M'$ by (2).

- If the last rule was $(cong_2)$, we proceed similarly to the last case.

- If the last rule was (ξ), then $M = \lambda x.N$ and $M' = \lambda x.N'$ for some N and N' such that $N \to_{\beta\eta} N'$. By induction hypothesis, $N \rhd N'$, which implies $M \rhd M'$ by (3).

(b) We prove this by induction on a derivation of $M \rhd M'$. We distinguish several cases, depending on the last rule used in the derivation.

- If the last rule was (1), then $M = M' = x$, and we are done because $x \twoheadrightarrow_{\beta\eta} x$.

- If the last rule was (2), then $M = PN$ and $M' = P'N'$, for some P, P', N, N' with $P \rhd P'$ and $N \rhd N'$. By induction hypothesis, $P \twoheadrightarrow_{\beta\eta} P'$ and $N \twoheadrightarrow_{\beta\eta} N'$. Since $\twoheadrightarrow_{\beta\eta}$ satisfies (*cong*), it follows that $PN \twoheadrightarrow_{\beta\eta} P'N'$, hence $M \twoheadrightarrow_{\beta\eta} M'$ as desired.

- If the last rule was (3), then $M = \lambda x.N$ and $M' = \lambda x.N'$, for some N, N' with $N \rhd N'$. By induction hypothesis, $N \twoheadrightarrow_{\beta\eta} N'$, hence $M = \lambda x.N \twoheadrightarrow_{\beta\eta} \lambda x.N' = M'$ by (ξ).

- If the last rule was (4), then $M = (\lambda x.Q)N$ and $M' = Q'[N'/x]$, for some Q, Q', N, N' with $Q \rhd Q'$ and $N \rhd N'$. By induction hypothesis, $Q \twoheadrightarrow_{\beta\eta} Q'$ and $N \twoheadrightarrow_{\beta\eta} N'$. Therefore $M = (\lambda x.Q)N \twoheadrightarrow_{\beta\eta} (\lambda x.Q')N' \to_{\beta\eta} Q'[N'/x] = M'$, as desired.

- If the last rule was (5), then $M = \lambda x.Px$ and $M' = P'$, for some P, P' with $P \rhd P'$, and $x \notin FV(P)$. By induction hypothesis, $P \twoheadrightarrow_{\beta\eta} P'$, hence $M = \lambda x.Px \to_{\beta\eta} P \twoheadrightarrow_{\beta\eta} P' = M'$, as desired.

(c) This follows directly from (a) and (b). Let us write R^* for the reflexive transitive closure of a relation R. By (a), we have $\to_{\beta\eta} \subseteq \rhd$, hence $\twoheadrightarrow_{\beta\eta} = \to_{\beta\eta}{}^* \subseteq \rhd^*$. By (b), we have $\rhd \subseteq \twoheadrightarrow_{\beta\eta}$, hence $\rhd^* \subseteq \twoheadrightarrow_{\beta\eta}{}^* = \twoheadrightarrow_{\beta\eta}$. It follows that $\rhd^* = \twoheadrightarrow_{\beta\eta}$. □

We will soon prove that $\rhd$ satisfies the diamond property. Note that together with Lemma 4.6(c), this will immediately imply that $\twoheadrightarrow_{\beta\eta}$ satisfies the Church-Rosser property.

Lemma 4.7 (Substitution). *If $M \rhd M'$ and $U \rhd U'$, then $M[U/y] \rhd M'[U'/y]$.*

Proof. We assume without loss of generality that any bound variables of M are different from y and from the free variables of U. The claim is now proved by induction on derivations of $M \rhd M'$. We distinguish several cases, depending on the last rule used in the derivation:

- If the last rule was (1), then $M = M' = x$, for some variable x. If $x = y$, then $M[U/y] = U \rhd U' = M'[U'/y]$. If $x \neq y$, then by (1), $M[U/y] = x \rhd x = M'[U'/y]$.

- If the last rule was (2), then $M = PN$ and $M' = P'N'$, for some P, P', N, N' with $P \rhd P'$ and $N \rhd N'$. By induction hypothesis, $P[U/y] \rhd P'[U'/y]$ and $N[U/y] \rhd N'[U'/y]$, hence by (2), $M[U/y] = P[U/y]N[U/y] \rhd P'[U'/y]N'[U'/y] = M'[U'/y]$.

- If the last rule was (3), then $M = \lambda x.N$ and $M' = \lambda x.N'$, for some N, N' with $N \rhd N'$. By induction hypothesis, $N[U/y] \rhd N'[U'/y]$, hence by (3) $M[U/y] = \lambda x.N[U/y] \rhd \lambda x.N'[U'/y] = M'[U'/y]$.

- If the last rule was (4), then $M = (\lambda x.Q)N$ and $M' = Q'[N'/x]$, for some Q, Q', N, N' with $Q \rhd Q'$ and $N \rhd N'$. By induction hypothesis, $Q[U/y] \rhd Q'[U'/y]$ and $N[U/y] \rhd N'[U'/y]$, hence by (4), $(\lambda x.Q[U/y])N[U/y] \rhd Q'[U'/y][N'[U'/y]/x] = Q'[N'/x][U'/y]$. Thus $M[U/y] \rhd M'[U'/y]$.

- If the last rule was (5), then $M = \lambda x.Px$ and $M' = P'$, for some P, P' with $P \rhd P'$, and $x \not\in FV(P)$. By induction hypothesis, $P[U/y] \rhd P'[U'/y]$, hence by (5), $M[U/y] = \lambda x.P[U/y]x \rhd P'[U'/y] = M'[U'/y]$. □

A more conceptual way of looking at this proof is the following: consider any derivation of $M \rhd M'$ from axioms (1)–(5). In this derivation, replace any axiom $y \rhd y$ by $U \rhd U'$, and propagate the changes (i.e., replace y by U on the left-hand-side, and by U' on the right-hand-side of any $\rhd$). The result is a derivation of $M[U/y] \rhd M'[U'/y]$. (The formal proof that the result of this replacement is indeed a valid derivation requires an induction, and this is the reason why the proof of the substitution lemma is so long).

Our next goal is to prove that $\rhd$ satisfies the diamond property. Before proving this, we first define the *maximal parallel one-step reduct* M^* of a term M as follows:

1. $x^* = x$, for a variable.
2. $(PN)^* = P^*N^*$, if PN is not a β-redex.
3. $((\lambda x.Q)N)^* = Q^*[N^*/x]$.
4. $(\lambda x.N)^* = \lambda x.N^*$, if $\lambda x.N$ is not an η-redex.
5. $(\lambda x.Px)^* = P^*$, if $x \not\in FV(P)$.

Note that M^* depends only on M. The following lemma implies the diamond property for $\rhd$.

Lemma 4.8 (Maximal parallel one-step reductions). *Whenever $M \rhd M'$, then $M' \rhd M^*$.*

Proof. By induction on the size of M. We distinguish five cases, depending on the last rule used in the derivation of $M \rhd M'$. As usual, we assume that all bound variables have been renamed to avoid clashes.

- If the last rule was (1), then $M = M' = x$, also $M^* = x$, and we are done.

- If the last rule was (2), then $M = PN$ and $M' = P'N'$, where $P \rhd P'$ and $N \rhd N'$. By induction hypothesis $P' \rhd P^*$ and $N' \rhd N^*$. Two cases:

  - If PN is not a β-redex, then $M^* = P^*N^*$. Thus $M' = P'N' \rhd P^*N^* = M^*$ by (2), and we are done.

- If PN is a β-redex, say $P = \lambda x.Q$, then $M^* = Q^*[N^*/x]$. We distinguish two subcases, depending on the last rule used in the derivation of $P \rhd P'$:

  * If the last rule was (3), then $P' = \lambda x.Q'$, where $Q \rhd Q'$. By induction hypothesis $Q' \rhd Q^*$, and with $N' \rhd N^*$, it follows that $M' = (\lambda x.Q')N' \rhd Q^*[N^*/x] = M^*$ by (4).
  * If the last rule was (5), then $P = \lambda x.Rx$ and $P' = R'$, where $x \notin FV(R)$ and $R \rhd R'$. Consider the term $Q = Rx$. Since $Rx \rhd R'x$, and Rx is a subterm of M, by induction hypothesis $R'x \rhd (Rx)^*$. By the substitution lemma, $M' = R'N' = (R'x)[N'/x] \rhd (Rx)^*[N^*/x] = M^*$.

- If the last rule was (3), then $M = \lambda x.N$ and $M' = \lambda x.N'$, where $N \rhd N'$. Two cases:

  - If M is not an η-redex, then $M^* = \lambda x.N^*$. By induction hypothesis, $N' \rhd N^*$, hence $M' \rhd M^*$ by (3).
  - If M is an η-redex, then $N = Px$, where $x \notin FV(P)$. In this case, $M^* = P^*$. We distinguish two subcases, depending on the last rule used in the derivation of $N \rhd N'$:

    * If the last rule was (2), then $N' = P'x$, where $P \rhd P'$. By induction hypothesis $P' \rhd P^*$. Hence $M' = \lambda x.P'x \rhd P^* = M^*$ by (5).
    * If the last rule was (4), then $P = \lambda y.Q$ and $N' = Q'[x/y]$, where $Q \rhd Q'$. Then $M' = \lambda x.Q'[x/y] = \lambda y.Q'$ (note $x \notin FV(Q')$). But $P \rhd \lambda y.Q'$, hence by induction hypothesis, $\lambda y.Q' \rhd P^* = M^*$.

- If the last rule was (4), then $M = (\lambda x.Q)N$ and $M' = Q'[N'/x]$, where $Q \rhd Q'$ and $N \rhd N'$. Then $M^* = Q^*[N^*/x]$, and $M' \rhd M^*$ by the substitution lemma.

- If the last rule was (5), then $M = \lambda x.Px$ and $M' = P'$, where $P \rhd P'$ and $x \notin FV(P)$. Then $M^* = P^*$. By induction hypothesis, $P' \rhd P^*$, hence $M' \rhd M^*$. □

The previous lemma immediately implies the diamond property for $\rhd$:

Lemma 4.9 (Diamond property for $\rhd$). *If $M \rhd N$ and $M \rhd P$, then there exists Z such that $N \rhd Z$ and $P \rhd Z$.*

Proof. Take $Z = M^*$. □

Finally, we have a proof of the Church-Rosser Theorem:

Proof of Theorem 4.2: Since $\triangleright$ satisfies the diamond property, it follows that its reflexive transitive closure $\triangleright^*$ also satisfies the diamond property, as shown in Figure 3. But $\triangleright^*$ is the same as $\twoheadrightarrow_{\beta\eta}$ by Lemma 4.6(c), and the diamond property for $\twoheadrightarrow_{\beta\eta}$ is just the Church-Rosser property for $\to_{\beta\eta}$. □

4.5 Exercises

Exercise 13. Give a detailed proof that property (c) from Section 4.3 implies property (a).

Exercise 14. Prove that $M \triangleright M$, for all terms M.

Exercise 15. Without using Lemma 4.8, prove that $M \triangleright M^*$ for all terms M.

Exercise 16. Let $\Omega = (\lambda x.xx)(\lambda x.xx)$. Prove that $\Omega \neq_{\beta\eta} \Omega\Omega$.

Exercise 17. What changes have to be made to Section 4.4 to get a proof of the Church-Rosser Theorem for $\to_\beta$, instead of $\to_{\beta\eta}$?

Exercise 18. Recall the properties (a)–(c) of binary relations $\to$ that were discussed in Section 4.3. Consider the following similar property, which is sometimes called the "strip property":

(d)

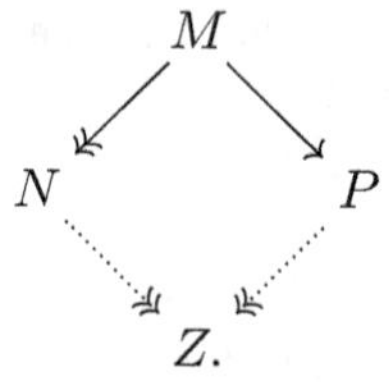

Does (d) imply (a)? Does (b) imply (d)? In each case, give either a proof or a counterexample.

Exercise 19. To every lambda term M, we may associate a directed graph (with possibly multiple edges and loops) $\mathcal{G}(M)$ as follows: (i) the vertices are terms N such that $M \twoheadrightarrow_\beta N$, i.e., all the terms that M can β-reduce to; (ii) the edges are given by a single-step β-reduction. Note that the same term may have two (or more) reductions coming from different redexes; each such reduction is a separate edge. For example, let $I = \lambda x.x$. Let $M = I(Ix)$. Then

$$\mathcal{G}(M) = I(Ix) \rightrightarrows Ix \longrightarrow x\ .$$

Note that there are two separate edges from $I(Ix)$ to Ix. We also sometimes write bullets instead of terms, to get $\bullet \rightrightarrows \bullet \longrightarrow \bullet$. As another example, let $\Omega = (\lambda x.xx)(\lambda x.xx)$. Then

$$\mathcal{G}(\Omega) = \bullet \circlearrowleft\ .$$

(a) Let $M = (\lambda x.I(xx))(\lambda x.xx)$. Find $\mathcal{G}(M)$.

(b) For each of the following graphs, find a term M such that $\mathcal{G}(M)$ is the given graph, or explain why no such term exists. (Note: the "starting" vertex need not always be the leftmost vertex in the picture). Warning: some of these terms are tricky to find!

(i)

(ii)

(iii)

(iv)

(v)

(vi)

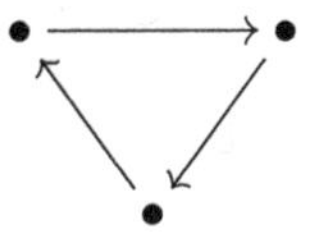

(vii)

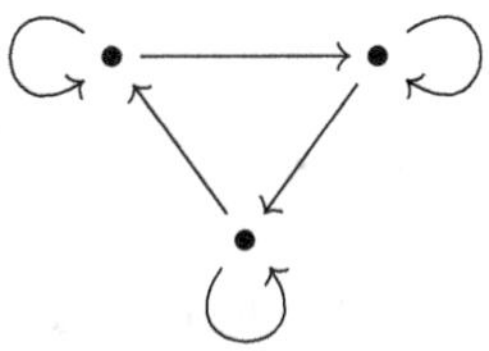

5 Combinatory algebras

To give a model of the lambda calculus means to provide a mathematical space in which the axioms of lambda calculus are satisfied. This usually means that the elements of the space can be understood as functions, and that certain functions can be understood as elements.

Naïvely, one might try to construct a model of lambda calculus by finding a set X such that X is in bijective correspondence with the set X^X of *all* functions from X to X. This, however, is impossible: for cardinality reasons, the equation $X \cong X^X$ has no solutions except for a one-element set $X = 1$. To see this, first note that the empty set $\emptyset$ is not a solution. Also, suppose X is a solution with $|X| \geqslant 2$. Then $|X^X| \geqslant |2^X|$, but by Cantor's argument, $|2^X| > |X|$, hence X^X is of greater cardinality than X, contradicting $X \cong X^X$.

There are two main strategies for constructing models of the lambda calculus, and both involve a restriction on the class of functions to make it smaller. The first approach, which will be discussed in this section, uses *algebra*, and the essential idea is to replace the set X^X of all function by a smaller, and suitably defined set of *polynomials*. The second approach is to equip the set X with additional structure (such as topology, ordered structure, etc), and to replace X^X by a set of structure-preserving functions (for example, continuous functions, monotone functions, etc).

5.1 Applicative structures

Definition. An *applicative structure* $(\mathbf{A}, \cdot)$ is a set $\mathbf{A}$ together with a binary operation "$\cdot$".

Note that there are no further assumptions; in particular, we do *not* assume that application is an associative operation. We write ab for $a \cdot b$, and as in the lambda calculus, we follow the convention of left associativity, i.e., we write abc for $(ab)c$.

Definition. Let $(\mathbf{A}, \cdot)$ be an applicative structure. A *polynomial* in a set of variables $x_1, \ldots, x_n$ and with coefficients in $\mathbf{A}$ is a formal expression built from variables and elements of $\mathbf{A}$ by means of the application operation. In other words, the set of polynomials is given by the following grammar:

$$t, s \quad ::= \quad x \mid a \mid ts,$$

where x ranges over variables and a ranges over the elements of $\mathbf{A}$. We write $\mathbf{A}\{x_1, \ldots, x_n\}$ for the set of polynomials in variables $x_1, \ldots, x_n$ with coefficients in $\mathbf{A}$.

Here are some examples of polynomials in the variables x, y, z, where $a, b \in \mathbf{A}$:

$$x, \qquad xy, \qquad axx, \qquad (x(y(zb)))(ax).$$

If $t(x_1, \ldots, x_n)$ is a polynomial in the indicated variables, and $b_1, \ldots, b_n$ are elements of $\mathbf{A}$, then we can evaluate the polynomial at the given elements: the evaluation $t(b_1, \ldots, b_n)$ is the element of $\mathbf{A}$ obtained by "plugging" $x_i = b_i$ into the polynomial, for $i = 1, \ldots, n$, and evaluating the resulting expression in $\mathbf{A}$. Note that in this way, every polynomial t in n variables can be understood as a *function* from $\mathbf{A}^n \to \mathbf{A}$. This is very similar to the usual polynomials in algebra, which can also either be understood as formal expressions or as functions.

If $t(x_1, \ldots, x_n)$ and $s(x_1, \ldots, x_n)$ are two polynomials with coefficients in $\mathbf{A}$, we say that the equation $t(x_1, \ldots, x_n) = s(x_1, \ldots, x_n)$ *holds* in $\mathbf{A}$ if for all $b_1, \ldots, b_n \in \mathbf{A}$, $t(b_1, \ldots, b_n) = s(b_1, \ldots, b_n)$.

5.2 Combinatory completeness

Definition (Combinatory completeness). An applicative structure $(\mathbf{A}, \cdot)$ is *combinatorially complete* if for every polynomial $t(x_1, \ldots, x_n)$ of $n \geqslant 0$ variables, there exists some element $a \in \mathbf{A}$ such that

$$ax_1 \ldots x_n = t(x_1, \ldots, x_n)$$

holds in $\mathbf{A}$.

In other words, combinatory completeness means that every polynomial *function* $t(x_1, \ldots, x_n)$ can be represented (in curried form) by some *element* of $\mathbf{A}$. We are therefore setting up a correspondence between functions and elements as discussed in the introduction of this section.

Note that we do not require the element a to be unique in the definition of combinatory completeness. This means that we are dealing with an intensional view of functions, where a given function might in general have several different names (but see the discussion of extensionality in Section 5.6).

The following theorem characterizes combinatory completeness in terms of a much simpler algebraic condition.

Theorem 5.1. *An applicative structure* $(\mathbf{A}, \cdot)$ *is combinatorially complete if and only if there exist two elements* $s, k \in \mathbf{A}$*, such that the following equations are satisfied for all* $x, y, z \in \mathbf{A}$*:*

$$\begin{array}{ll} (1) & sxyz = (xz)(yz) \\ (2) & kxy = x \end{array}$$

Example 5.2. Before we prove this theorem, let us look at a few examples.

(a) The identity function. Can we find an element $i \in \mathbf{A}$ such that $ix = x$ for all x? Yes, indeed, we can let $i = skk$. We check that for all x, $skkx = (kx)(kx) = x$.

(b) The boolean "true". Can we find an element $\mathbf{T}$ such that for all x, y, $\mathbf{T}xy = x$? Yes, this is easy: $\mathbf{T} = k$.

(c) The boolean "false". Can we find $\mathbf{F}$ such that $\mathbf{F}xy = y$? Yes, what we need is $\mathbf{F}x = i$. Therefore a solution is $\mathbf{F} = ki$. And indeed, for all y, we have $kixy = iy = y$.

(d) Find a function f such that $fx = xx$ for all x. Solution: let $f = sii$. Then $siix = (ix)(ix) = xx$.

Proof of Theorem 5.1: The "only if" direction is trivial. If $\mathbf{A}$ is combinatorially complete, then consider the polynomial $t(x, y, z) = (xz)(yz)$. By combinatory completeness, there exists some $s \in \mathbf{A}$ with $sxyz = t(x, y, z)$, and similarly for k.

We therefore have to prove the "if" direction. Recall that $\mathbf{A}\{x_1,\dots,x_n\}$ is the set of polynomials with variables $x_1,\dots,x_n$. Now for each polynomial $t \in \mathbf{A}\{x, y_1,\dots,y_n\}$ in $n+1$ variables, we will define a new polynomial $\lambda^* x.t \in \mathbf{A}\{y_1,\dots,y_n\}$ in n variables, as follows by recursion on t:

$$
\begin{array}{lcll}
\lambda^* x.x & := & i, & \\
\lambda^* x.y_i & := & k y_i & \text{where } y_i \neq x \text{ is a variable,} \\
\lambda^* x.a & := & k a & \text{where } a \in \mathbf{A}, \\
\lambda^* x.pq & := & s(\lambda^* x.p)(\lambda^* x.q). &
\end{array}
$$

We claim that for all t, the equation $(\lambda^* x.t)x = t$ holds in $\mathbf{A}$. Indeed, this is easily proved by induction on t, using the definition of λ^*:

$$
\begin{array}{lcl}
(\lambda^* x.x)x & = & ix = x, \\
(\lambda^* x.y_i)x & = & k y_i x = y_i, \\
(\lambda^* x.a)x & = & kax = a, \\
(\lambda^* x.pq)x & = & s(\lambda^* x.p)(\lambda^* x.q)x = ((\lambda^* x.p)x)((\lambda^* x.q)x) = pq.
\end{array}
$$

Note that the last case uses the induction hypothesis for p and q.

Finally, to prove the theorem, assume that $\mathbf{A}$ has elements s, k satisfying equations (1) and (2), and consider a polynomial $t \in \mathbf{A}\{x_1,\dots,x_n\}$. We must show that there exists $a \in \mathbf{A}$ such that $a x_1 \dots x_n = t$ holds in $\mathbf{A}$. We let

$$a = \lambda^* x_1. \dots . \lambda^* x_n.t.$$

Note that a is a polynomial in 0 variables, which we may consider as an element of $\mathbf{A}$. Then from the previous claim, it follows that

$$
\begin{array}{lcl}
a x_1 \dots x_n & = & (\lambda^* x_1.\lambda^* x_2. \dots . \lambda^* x_n.t) x_1 x_2 \dots x_n \\
 & = & (\lambda^* x_2. \dots . \lambda^* x_n.t) x_2 \dots x_n \\
 & = & \dots \\
 & = & (\lambda^* x_n.t) x_n \\
 & = & t
\end{array}
$$

holds in $\mathbf{A}$. □

5.3 Combinatory algebras

By Theorem 5.1, combinatory completeness is equivalent to the existence of the s and k operators. We enshrine this in the following definition:

Definition (Combinatory algebra). A *combinatory algebra* $(\mathbf{A}, \cdot, s, k)$ is an applicative structure $(\mathbf{A}, \cdot)$ together with elements $s, k \in \mathbf{A}$, satisfying the following two axioms:

$$
\begin{array}{ll}
(1) & sxyz = (xz)(yz) \\
(2) & kxy = x
\end{array}
$$

Remark 5.3. The operation λ^*, defined in the proof of Theorem 5.1, is defined on the polynomials of any combinatory algebra. It is called the *derived lambda abstractor*, and it satisfies the law of β-equivalence, i.e., $(\lambda^* x.t)b = t[b/x]$, for all $b \in \mathbf{A}$.

Finding actual examples of combinatory algebras is not so easy. Here are some examples:

Example 5.4. The one-element set $\mathbf{A} = \{*\}$, with $* \cdot * = *$, $s = *$, and $k = *$, is a combinatory algebra. It is called the *trivial* combinatory algebra.

Example 5.5. Recall that Λ is the set of lambda terms. Let $\mathbf{A} = \Lambda/{=_\beta}$, the set of lambda terms modulo β-equivalence. Define $M \cdot N = MN$, $S = \lambda xyz.(xz)(yz)$, and $K = \lambda xy.x$. Then $(\Lambda, \cdot, S, K)$ is a combinatory algebra. Also note that, by Corollary 4.5, this algebra is non-trivial, i.e., it has more than one element.

Similar examples are obtained by replacing $=_\beta$ by $=_{\beta\eta}$, and/or replacing Λ by the set Λ_0 of closed terms.

Example 5.6. We construct a combinatory algebra of SK-terms as follows. Let V be a given set of variables. The set $\mathfrak{C}$ of *terms* of combinatory logic is given by the grammar:

$$A, B \quad ::= \quad x \mid \mathbf{S} \mid \mathbf{K} \mid AB,$$

where x ranges over the elements of V.

On $\mathfrak{C}$, we define combinatory equivalence $=_c$ as the smallest equivalence relation satisfying $\mathbf{S}ABC =_c (AC)(BC)$, $\mathbf{K}AB =_c A$, and the rules ($cong_1$) and ($cong_2$) (see page 11). Then the set $\mathfrak{C}/{=_c}$ is a combinatory algebra (called the *free* combinatory algebra generated by V, or the *term algebra*). You will prove in Exercise 20 that it is non-trivial.

Exercise 20. On the set $\mathfrak{C}$ of combinatory terms, define a notion of *single-step reduction* by the following laws:

$$\mathbf{S}ABC \to_c (AC)(BC),$$
$$\mathbf{K}AB \to_c A,$$

together with the usual rules ($cong_1$) and ($cong_2$) (see page 11). As in lambda calculus, we call a term a *normal form* if it cannot be reduced. Prove that the reduction $\to_c$ satisfies the Church-Rosser property. (Hint: similarly to the lambda calculus, first define a suitable parallel one-step reduction $\triangleright$ whose reflexive transitive closure is that of $\to_c$. Then show that it satisfies the diamond property.)

Corollary 5.7. *It immediately follows from the Church-Rosser Theorem for combinatory logic (Exercise 20) that two normal forms are $=_c$-equivalent if and only if they are equal.*

5.4 The failure of soundness for combinatory algebras

A combinatory algebra is almost a model of the lambda calculus. Indeed, given a combinatory algebra $\mathbf{A}$, we can interpret any lambda term as follows. To each

(*refl*)	$\dfrac{}{M = M}$	(*cong*)	$\dfrac{M = M' \qquad N = N'}{MN = M'N'}$
(*symm*)	$\dfrac{M = N}{N = M}$	(ξ)	$\dfrac{M = M'}{\lambda x.M = \lambda x.M'}$
(*trans*)	$\dfrac{M = N \qquad N = P}{M = P}$	(β)	$\dfrac{}{(\lambda x.M)N = M[N/x]}$

Table 2: The rules for β-equivalence

term M with free variables among $x_1, \ldots, x_n$, we recursively associate a polynomial $[\![M]\!] \in \mathbf{A}\{x_1, \ldots, x_n\}$:

$$\begin{aligned} [\![x]\!] &:= x, \\ [\![NP]\!] &:= [\![N]\!][\![P]\!], \\ [\![\lambda x.M]\!] &:= \lambda^* x.[\![M]\!]. \end{aligned}$$

Notice that this definition is almost the identity function, except that we have replaced the ordinary lambda abstractor of lambda calculus by the derived lambda abstractor of combinatory logic. The result is a polynomial in $\mathbf{A}\{x_1, \ldots, x_n\}$. In the particular case where M is a closed term, we can regard $[\![M]\!]$ as an element of $\mathbf{A}$.

To be able to say that $\mathbf{A}$ is a "model" of the lambda calculus, we would like the following property to be true:

$$M =_\beta N \Rightarrow [\![M]\!] = [\![N]\!] \text{ holds in } \mathbf{A}.$$

This property is called *soundness* of the interpretation. Unfortunately, it is in general false for combinatory algebras, as the following example shows.

Example 5.8. Let $M = \lambda x.x$ and $N = \lambda x.(\lambda y.y)x$. Then clearly $M =_\beta N$. On the other hand,

$$\begin{aligned} [\![M]\!] &= \lambda^* x.x = i, \\ [\![N]\!] &= \lambda^* x.(\lambda^* y.y)x = \lambda^* x.ix = s(ki)i. \end{aligned}$$

It follows from Exercise 20 and Corollary 5.7 that the equation $i = s(ki)i$ does not hold in the combinatory algebra $\mathfrak{C}/{=_c}$. In other words, the interpretation is not sound.

Let us analyze the failure of the soundness property further. Recall that β-equivalence is the smallest equivalence relation on lambda terms satisfying the six rules in Table 2.

If we define a relation $\sim$ on lambda terms by

$$M \sim N \qquad \Longleftrightarrow \qquad [\![M]\!] = [\![N]\!] \text{ holds in } \mathbf{A},$$

then we may ask which of the six rules of Table 2 the relation $\sim$ satisfies. Clearly, not all six rules can be satisfied, or else we would have $M =_\beta N \Rightarrow M \sim N \Rightarrow [\![M]\!] = [\![N]\!]$, i.e., the model would be sound.

Clearly, $\sim$ is an equivalence relation, and therefore satisfies (*refl*), (*symm*), and (*trans*). Also, (*cong*) is satisfied, because whenever p, q, p', q' are polynomials such that $p = p'$ and $q = q'$ holds in **A**, then clearly $pq = p'q'$ holds in **A** as well. Finally, we know from Remark 5.3 that the rule (β) is satisfied.

So the rule that fails is the (ξ) rule. Indeed, Example 5.8 illustrates this. Note that $x \sim (\lambda y.y)x$ (from the proof of Theorem 5.1), but $\lambda x.x \not\sim \lambda x.(\lambda y.y)x$, and therefore the ($\xi$) rule is violated.

5.5 Lambda algebras

A lambda algebra is, by definition, a combinatory algebra that is a sound model of lambda calculus, and in which s and k have their expected meanings.

Definition (Lambda algebra)**.** A *lambda algebra* is a combinatory algebra **A** satisfying the following properties:

$$\begin{array}{ll} (\forall M, N \in \Lambda)\ M =_\beta N \ \Rightarrow\ [\![M]\!] = [\![N]\!] & (\textit{soundness}), \\ s = \lambda^* x.\lambda^* y.\lambda^* z.(xz)(yz) & (\textit{s-derived}), \\ k = \lambda^* x.\lambda^* y.x & (\textit{k-derived}). \end{array}$$

The purpose of the remainder of this section is to give an axiomatic description of lambda algebras.

Lemma 5.9. *Recall that Λ_0 is the set of closed lambda terms, i.e., lambda terms without free variables. Soundness is equivalent to the following:*

$$(\forall M, N \in \Lambda_0)\ M =_\beta N \ \Rightarrow\ [\![M]\!] = [\![N]\!] \quad \textit{(closed soundness)}$$

Proof. Clearly soundness implies closed soundness. For the converse, assume closed soundness and let $M, N \in \Lambda$ with $M =_\beta N$. Let $FV(M) \cup FV(N) = \{x_1, \ldots, x_n\}$. Then

$$\begin{array}{lll} M =_\beta N & & \\ \Rightarrow & \lambda x_1 \ldots x_n.M =_\beta \lambda x_1 \ldots x_n.N & \text{by } (\xi) \\ \Rightarrow & [\![\lambda x_1 \ldots x_n.M]\!] = [\![\lambda x_1 \ldots x_n.N]\!] & \text{by closed soundness} \\ \Rightarrow & \lambda^* x_1 \ldots x_n.[\![M]\!] = \lambda^* x_1 \ldots x_n.[\![N]\!] & \text{by def. of } [\![-]\!] \\ \Rightarrow & (\lambda^* x_1 \ldots x_n.[\![M]\!])x_1 \ldots x_n & \\ & \quad = (\lambda^* x_1 \ldots x_n.[\![N]\!])x_1 \ldots x_n & \\ \Rightarrow & [\![M]\!] = [\![N]\!] & \text{by proof of Thm 5.1} \end{array}$$

This proves soundness. □

Definition (Translations between combinatory logic and lambda calculus). Let $A \in \mathfrak{C}$ be a combinatory term (see Example 5.6). We define its translation to lambda calculus in the obvious way: the translation A_λ is given recursively by:

$$\begin{array}{lcl} \mathbf{S}_\lambda & = & \lambda xyz.(xz)(yz), \\ \mathbf{K}_\lambda & = & \lambda xy.x, \\ x_\lambda & = & x, \\ (AB)_\lambda & = & A_\lambda B_\lambda. \end{array}$$

Conversely, given a lambda term $M \in \Lambda$, we recursively define its translation M_c to combinatory logic like this:

$$\begin{array}{lcl} x_c & = & x, \\ (MN)_c & = & M_c N_c, \\ (\lambda x.M)_c & = & \lambda^* x.(M_c). \end{array}$$

Lemma 5.10. *For all lambda terms M, $(M_c)_\lambda =_\beta M$.*

Lemma 5.11. *Let $\mathbf{A}$ be a combinatory algebra satisfying $k = \lambda^* x.\lambda^* y.x$ and $s = \lambda^* x.\lambda^* y.\lambda^* z.(xz)(yz)$. Then for all combinatory terms A, $(A_\lambda)_c = A$ holds in $\mathbf{A}$.*

Exercise 21. Prove Lemmas 5.10 and 5.11.

Let $\mathfrak{C}_0$ be the set of *closed* combinatory terms. The following is our first useful characterization of lambda calculus.

Lemma 5.12. *Let $\mathbf{A}$ be a combinatory algebra. Then $\mathbf{A}$ is a lambda algebra if and only if it satisfies the following property:*

$$(\forall A, B \in \mathfrak{C}_0)\ A_\lambda =_\beta B_\lambda\ \Rightarrow\ A = B \text{ holds in } \mathbf{A}. \quad \textit{(alt-soundness)}$$

Proof. First, assume that $\mathbf{A}$ satisfies (*alt-soundness*). To prove closed soundness, let M, N be closed lambda terms with $M =_\beta N$. Then $(M_c)_\lambda =_\beta M =_\beta N =_\beta (N_c)_\lambda$, hence by (*alt-soundness*), $M_c = N_c$ holds in $\mathbf{A}$. But this is the definition of $[\![M]\!] = [\![N]\!]$.

To prove (*k-derived*), note that

$$\begin{array}{rlll} \mathbf{K}_\lambda & = & (\lambda x.\lambda y.x) & \text{by definition of } (-)_\lambda \\ & =_\beta & ((\lambda x.\lambda y.x)_c)_\lambda & \text{by Lemma 5.10} \\ & = & (\lambda^* x.\lambda^* y.x)_\lambda & \text{by definition of } (-)_c. \end{array}$$

Hence, by (*alt-soundness*), it follows that $\mathbf{K} = (\lambda^* x.\lambda^* y.x)$ holds in $\mathbf{A}$. Similarly for (*s-derived*).

Conversely, assume that $\mathbf{A}$ is a lambda algebra. Let $A, B \in \mathfrak{C}_0$ and assume $A_\lambda =_\beta B_\lambda$. By soundness, $[\![A_\lambda]\!] = [\![B_\lambda]\!]$. By definition of the interpretation, $(A_\lambda)_c = (B_\lambda)_c$ holds in $\mathbf{A}$. But by (*s-derived*), (*k-derived*), and Lemma 5.11, $A = (A_\lambda)_c = (B_\lambda)_c = B$ holds in $\mathbf{A}$, proving (*alt-soundness*). □

Definition (Homomorphism). Let $(\mathbf{A}, \cdot_{\mathbf{A}}, s_{\mathbf{A}}, k_{\mathbf{A}})$, $(\mathbf{B}, \cdot_{\mathbf{B}}, s_{\mathbf{B}}, k_{\mathbf{B}})$ be combinatory algebras. A *homomorphism* of combinatory algebras is a function $\varphi : \mathbf{A} \to \mathbf{B}$ such that $\varphi(s_{\mathbf{A}}) = s_{\mathbf{B}}$, $\varphi(k_{\mathbf{A}}) = k_{\mathbf{B}}$, and for all $a, b \in \mathbf{A}$, $\varphi(a \cdot_{\mathbf{A}} b) = \varphi(a) \cdot_{\mathbf{B}} \varphi(b)$.

Any given homomorphism $\varphi : \mathbf{A} \to \mathbf{B}$ can be extended to polynomials in the obvious way: we define $\hat{\varphi} : \mathbf{A}\{x_1, \ldots, x_n\} \to \mathbf{B}\{x_1, \ldots, x_n\}$ by

$$\begin{array}{ll} \hat{\varphi}(a) = \varphi(a) & \text{for } a \in \mathbf{A}, \\ \hat{\varphi}(x) = x & \text{if } x \in \{x_1, \ldots, x_n\}, \\ \hat{\varphi}(pq) = \hat{\varphi}(p)\hat{\varphi}(q). & \end{array}$$

Example 5.13. If $\varphi(a) = a'$ and $\varphi(b) = b'$, then $\hat{\varphi}((ax)(by)) = (a'x)(b'y)$.

The following is the main technical concept needed in the characterization of lambda algebras. We say that an equation *holds absolutely* if it holds in $\mathbf{A}$ and in any homomorphic image of $\mathbf{A}$. If an equation holds only in the previous sense, then we sometimes say it holds *locally*.

Definition (Absolute equation). Let $p, q \in \mathbf{A}\{x_1, \ldots, x_n\}$ be two polynomials with coefficients in $\mathbf{A}$. We say that the equation $p = q$ *holds absolutely* in $\mathbf{A}$ if for all combinatory algebras $\mathbf{B}$ and all homomorphisms $\varphi : \mathbf{A} \to \mathbf{B}$, $\hat{\varphi}(p) = \hat{\varphi}(q)$ holds in $\mathbf{B}$. If an equation holds absolutely, we write $p =_{\text{abs}} q$.

We can now state the main theorem characterizing lambda algebras. Let $\mathbf{1} = s(ki)$.

Theorem 5.14. *Let* $\mathbf{A}$ *be a combinatory algebra. Then the following are equivalent:*

1. $\mathbf{A}$ *is a lambda algebra,*

2. $\mathbf{A}$ *satisfies (alt-soundness),*

3. *for all* $A, B \in \mathfrak{C}$ *such that* $A_\lambda =_\beta B_\lambda$, *the equation* $A = B$ *holds absolutely in* $\mathbf{A}$,

4. $\mathbf{A}$ *absolutely satisfies the nine axioms in Table 3,*

5. $\mathbf{A}$ *satisfies (s-derived) and (k-derived), and for all* $p, q \in \mathbf{A}\{y_1, \ldots, y_n\}$, *if* $px =_{\text{abs}} qx$ *then* $\mathbf{1}p =_{\text{abs}} \mathbf{1}q$,

6. $\mathbf{A}$ *satisfies (s-derived) and (k-derived), and for all* $p, q \in \mathbf{A}\{x, y_1, \ldots, y_n\}$, *if* $p =_{\text{abs}} q$ *then* $\lambda^* x.p =_{\text{abs}} \lambda^* y.q$.

The proof proceeds via $1 \Rightarrow 2 \Rightarrow 3 \Rightarrow 4 \Rightarrow 5 \Rightarrow 6 \Rightarrow 1$.

We have already proven $1 \Rightarrow 2$ in Lemma 5.12.

To prove $2 \Rightarrow 3$, let $FV(A) \cup FV(B) \subseteq \{x_1, \ldots, x_n\}$, and assume $A_\lambda =_\beta B_\lambda$. Then $\lambda x_1 \ldots x_n.(A_\lambda) =_\beta \lambda x_1 \ldots x_n.(B_\lambda)$, hence $(\lambda^* x_1 \ldots x_n.A)_\lambda =_\beta$

$$\begin{array}{lrcl}
(a) & \mathbf{1}k & =_{\text{abs}} & k, \\
(b) & \mathbf{1}s & =_{\text{abs}} & s, \\
(c) & \mathbf{1}(kx) & =_{\text{abs}} & kx, \\
(d) & \mathbf{1}(sx) & =_{\text{abs}} & sx, \\
(e) & \mathbf{1}(sxy) & =_{\text{abs}} & sxy, \\
(f) & s(s(kk)x)y & =_{\text{abs}} & \mathbf{1}x, \\
(g) & s(s(s(ks)x)y)z & =_{\text{abs}} & s(sxz)(syz), \\
(h) & k(xy) & =_{\text{abs}} & s(kx)(ky), \\
(i) & s(kx)i & =_{\text{abs}} & \mathbf{1}x.
\end{array}$$

Table 3: An axiomatization of lambda algebras. Here $\mathbf{1} = s(ki)$.

$(\lambda^* x_1 \ldots x_n.B)_\lambda$ (why?). Since the latter terms are closed, it follows by the rule (*alt-soundness*) that $\lambda^* x_1 \ldots x_n.A = \lambda^* x_1 \ldots x_n.B$ holds in $\mathbf{A}$. Since closed equations are preserved by homomorphisms, the latter also holds in $\mathbf{B}$ for any homomorphism $\varphi : \mathbf{A} \to \mathbf{B}$. Finally, this implies that $A = B$ holds for any such $\mathbf{B}$, proving that $A = B$ holds absolutely in $\mathbf{A}$.

Exercise 22. Prove the implication $3 \Rightarrow 4$.

The implication $4 \Rightarrow 5$ is the most difficult part of the theorem. We first dispense with the easier part:

Exercise 23. Prove that the axioms from Table 3 imply (*s-derived*) and (*k-derived*).

The last part of $4 \Rightarrow 5$ needs the following lemma:

Lemma 5.15. *Suppose* $\mathbf{A}$ *satisfies the nine axioms from Table 3. Define a structure* $(\mathbf{B}, \bullet, S, K)$ *by:*

$$\begin{array}{l}
\mathbf{B} = \{a \in \mathbf{A} \mid a = \mathbf{1}a\}, \\
a \bullet b = sab, \\
S = ks, \\
K = kk.
\end{array}$$

Then $\mathbf{B}$ *is a well-defined combinatory algebra. Moreover, the function* $\varphi : \mathbf{A} \to \mathbf{B}$ *defined by* $\varphi(a) = ka$ *defines a homomorphism.*

Exercise 24. Prove Lemma 5.15.

To prove the implication $4 \Rightarrow 5$, assume $ax = bx$ holds absolutely in $\mathbf{A}$. Then $\hat{\varphi}(ax) = \hat{\varphi}(bx)$ holds in $\mathbf{B}$ by definition of "absolute". But $\hat{\varphi}(ax) = (\varphi a)x = s(ka)x$ and $\hat{\varphi}(bx) = (\varphi b)x = s(kb)x$. Therefore $s(ka)x = s(kb)x$ holds in $\mathbf{A}$. We plug in $x = i$ to get $s(ka)i = s(kb)i$. By axiom (i), $\mathbf{1}a = \mathbf{1}b$.

To prove $5 \Rightarrow 6$, assume $p =_{\text{abs}} q$. Then $(\lambda^* x.p)x =_{\text{abs}} p =_{\text{abs}} q =_{\text{abs}} (\lambda^* x.q)x$ by the proof of Theorem 5.1. Then by 5., $(\lambda^* x.p) =_{\text{abs}} (\lambda^* x.q)$.

Finally, to prove $6 \Rightarrow 1$, note that if 6 holds, then the absolute interpretation satisfies the ξ-rule, and therefore satisfies all the axioms of lambda calculus.

Exercise 25. Prove $6 \Rightarrow 1$.

Remark 5.16. The axioms in Table 3 are required to hold *absolutely*. They can be replaced by local axioms by prefacing each axiom with $\lambda^* xyz$. Note that this makes the axioms much longer.

5.6 Extensional combinatory algebras

Definition. An applicative structure $(\mathbf{A}, \cdot)$ is *extensional* if for all $a, b \in \mathbf{A}$, if $ac = bc$ holds for all $c \in \mathbf{A}$, then $a = b$.

Proposition 5.17. *In an extensional combinatory algebra, the* (η) *axioms is valid.*

Proof. By (β), $(\lambda^* x.Mx)c = Mc$ for all $c \in \mathbf{A}$. Therefore, by extensionality, $(\lambda^* x.Mx) = M$. □

Proposition 5.18. *In an extensional combinatory algebra, an equation holds locally if and only if it holds absolutely.*

Proof. Clearly, if an equation holds absolutely, then it holds locally. Conversely, assume the equation $p = q$ holds locally in $\mathbf{A}$. Let $x_1, \ldots, x_n$ be the variables occurring in the equation. By (β),

$$(\lambda^* x_1 \ldots x_n.p)x_1 \ldots x_n = (\lambda^* x_1 \ldots x_n.q)x_1 \ldots x_n$$

holds locally. By extensionality,

$$\lambda^* x_1 \ldots x_n.p = \lambda^* x_1 \ldots x_n.q$$

holds. Since this is a closed equation (no free variables), it automatically holds absolutely. This implies that $(\lambda^* x_1 \ldots x_n.p)x_1 \ldots x_n = (\lambda^* x_1 \ldots x_n.q)x_1 \ldots x_n$ holds absolutely, and finally, by (β) again, that $p = q$ holds absolutely. □

Proposition 5.19. *Every extensional combinatory algebra is a lambda algebra.*

Proof. By Theorem 5.14(6), it suffices to prove (*s-derived*), (*k-derived*) and the (ξ)-rule. Let $a, b, c \in \mathbf{A}$ be arbitrary. Then

$$(\lambda^* x.\lambda^* y.\lambda^* z.(xz)(yz))abc = (ac)(bc) = sabc$$

by (β) and definition of s. Applying extensionality three times (with respect to c, b, and a), we get

$$\lambda^* x.\lambda^* y.\lambda^* z.(xz)(yz) = s.$$

This proves (*s-derived*). The proof of (*k-derived*) is similar. Finally, to prove (ξ), assume that $p =_{\text{abs}} q$. Then by (β), $(\lambda^* x.p)c = (\lambda^* x.q)c$ for all $c \in \mathbf{A}$. By extensionality, $\lambda^* x.p = \lambda^* x.q$ holds. □

6 Simply-typed lambda calculus, propositional logic, and the Curry-Howard isomorphism

In the untyped lambda calculus, we spoke about functions without speaking about their domains and codomains. The domain and codomain of any function was the set of all lambda terms. We now introduce types into the lambda calculus, and thus a notion of domain and codomain for functions. The difference between types and sets is that types are *syntactic* objects, i.e., we can speak of types without having to speak of their elements. We can think of types as *names* for sets.

6.1 Simple types and simply-typed terms

We assume a set of basic types. We usually use the Greek letter ι ("iota") to denote a basic type. The set of simple types is given by the following BNF:

$$\text{Simple types:} \quad A, B ::= \iota \mid A \to B \mid A \times B \mid 1$$

The intended meaning of these types is as follows: base types are things like the type of integers or the type of booleans. The type $A \to B$ is the type of functions from A to B. The type $A \times B$ is the type of pairs $\langle x, y\rangle$, where x has type A and y has type B. The type 1 is a one-element type. You can think of 1 as an abridged version of the booleans, in which there is only one boolean instead of two. Or you can think of 1 as the "void" or "unit" type in many programming languages: the result type of a function that has no real result.

When we write types, we adopt the convention that $\times$ binds stronger than $\to$, and $\to$ associates to the right. Thus, $A \times B \to C$ is $(A \times B) \to C$, and $A \to B \to C$ is $A \to (B \to C)$.

The set of *raw typed lambda terms* is given by the following BNF:

$$\text{Raw terms:} \quad M, N ::= x \mid MN \mid \lambda x^A.M \mid \langle M, N\rangle \mid \pi_1 M \mid \pi_2 M \mid *$$

Unlike what we did in the untyped lambda calculus, we have added special syntax here for pairs. Specifically, $\langle M, N\rangle$ is a pair of terms, $\pi_i M$ is a projection, with the intention that $\pi_i\langle M_1, M_2\rangle = M_i$. Also, we have added a term $*$, which is the unique element of the type 1. One other change from the untyped lambda calculus is that we now write $\lambda x^A.M$ for a lambda abstraction to indicate that x has type A. However, we will sometimes omit the superscripts and write $\lambda x.M$ as before. The notions of free and bound variables and α-conversion are defined as for the untyped lambda calculus; again we identify α-equivalent terms.

We call the above terms the *raw* terms, because we have not yet imposed any typing discipline on these terms. To avoid meaningless terms such as $\langle M, N\rangle(P)$ or $\pi_1(\lambda x.M)$, we introduce *typing rules*.

We use the colon notation $M : A$ to mean "M is of type A". (Similar to the element notation in set theory). The typing rules are expressed in terms of *typing*

$$(var)\quad \frac{}{\Gamma, x{:}A \vdash x : A}$$

$$(app)\quad \frac{\Gamma \vdash M : A \to B \qquad \Gamma \vdash N : A}{\Gamma \vdash MN : B} \qquad (\pi_1)\quad \frac{\Gamma \vdash M : A \times B}{\Gamma \vdash \pi_1 M : A}$$

$$(abs)\quad \frac{\Gamma, x{:}A \vdash M : B}{\Gamma \vdash \lambda x^A.M : A \to B} \qquad (\pi_2)\quad \frac{\Gamma \vdash M : A \times B}{\Gamma \vdash \pi_2 M : B}$$

$$(pair)\quad \frac{\Gamma \vdash M : A \qquad \Gamma \vdash N : B}{\Gamma \vdash \langle M, N\rangle : A \times B} \qquad (*)\quad \frac{}{\Gamma \vdash * : 1}$$

Table 4: Typing rules for the simply-typed lambda calculus

judgments. A typing judgment is an expression of the form

$$x_1{:}A_1, x_2{:}A_2, \ldots, x_n{:}A_n \vdash M : A.$$

Its meaning is: "under the assumption that x_i is of type A_i, for $i = 1 \ldots n$, the term M is a well-typed term of type A." The free variables of M must be contained in $x_1, \ldots, x_n$. The idea is that in order to determine the type of M, we must make some assumptions about the type of its free variables. For instance, the term xy will have type B if $x{:}A \to B$ and $y{:}A$. Clearly, the type of xy depends on the type of its free variables.

A sequence of assumptions of the form $x_1{:}A_1, \ldots, x_n{:}A_n$, as in the left-hand-side of a typing judgment, is called a *typing context*. We always assume that no variable appears more than once in a typing context, and we allow typing contexts to be re-ordered implicitly. We often use the Greek letter Γ to stand for an arbitrary typing context, and we use the notations Γ, Γ' and $\Gamma, x{:}A$ to denote the concatenation of typing contexts, where it is always assumed that the sets of variables are disjoint.

The symbol $\vdash$, which appears in a typing judgment, is called the *turnstile* symbol. Its purpose is to separate the left-hand side from the right-hand side.

The typing rules for the simply-typed lambda calculus are shown in Table 4. The rule (*var*) is a tautology: under the assumption that x has type A, x has type A. The rule (*app*) states that a function of type $A \to B$ can be applied to an argument of type A to produce a result of type B. The rule (*abs*) states that if M is a term of type B with a free variable x of type A, then $\lambda x^A.M$ is a function of type $A \to B$. The other rules have similar interpretations.

Here is an example of a valid typing derivation:

$$\cfrac{\cfrac{\cfrac{\cfrac{}{x{:}A \to A, y{:}A \vdash x : A \to A} \qquad \cfrac{\cfrac{}{x{:}A \to A, y{:}A \vdash x : A \to A} \qquad \cfrac{}{x{:}A \to A, y{:}A \vdash y : A}}{x{:}A \to A, y{:}A \vdash xy : A}}{x{:}A \to A, y{:}A \vdash x(xy) : A}}{x{:}A \to A \vdash \lambda y^A.x(xy) : A \to A}}{\vdash \lambda x^{A \to A}.\lambda y^A.x(xy) : (A \to A) \to A \to A}$$

One important property of these typing rules is that there is precisely one rule for each kind of lambda term. Thus, when we construct typing derivations in a bottom-up fashion, there is always a unique choice of which rule to apply next. The only real choice we have is about which types to assign to variables.

Exercise 26. Give a typing derivation of each of the following typing judgments:

(a) $\vdash \lambda x^{(A\to A)\to B}.x(\lambda y^A.y) : ((A \to A) \to B) \to B$

(b) $\vdash \lambda x^{A\times B}.\langle \pi_2 x, \pi_1 x\rangle : (A \times B) \to (B \times A)$

Not all terms are typable. For instance, the terms $\pi_1(\lambda x.M)$ and $\langle M, N\rangle(P)$ cannot be assigned a type, and neither can the term $\lambda x.xx$. Here, by "assigning a type" we mean, assigning types to the free and bound variables such that the corresponding typing judgment is derivable. We say that a term is typable if it can be assigned a type.

Exercise 27. Show that neither of the three terms mentioned in the previous paragraph is typable.

Exercise 28. We said that we will identify α-equivalent terms. Show that this is actually necessary. In particular, show that if we didn't identify α-equivalent terms, there would be no valid derivation of the typing judgment

$$\vdash \lambda x^A.\lambda x^B.x : A \to B \to B.$$

Give a derivation of this typing judgment using the bound variable convention.

6.2 Connections to propositional logic

Consider the following types:

$$\begin{array}{ll}
(1) & (A \times B) \to A \\
(2) & A \to B \to (A \times B) \\
(3) & (A \to B) \to (B \to C) \to (A \to C) \\
(4) & A \to A \to A \\
(5) & ((A \to A) \to B) \to B \\
(6) & A \to (A \times B) \\
(7) & (A \to C) \to C
\end{array}$$

Let us ask, in each case, whether it is possible to find a closed term of the given type. We find the following terms:

$$\begin{array}{ll}
(1) & \lambda x^{A\times B}.\pi_1 x \\
(2) & \lambda x^A.\lambda y^B.\langle x, y\rangle \\
(3) & \lambda x^{A\to B}.\lambda y^{B\to C}.\lambda z^A.y(xz) \\
(4) & \lambda x^A.\lambda y^A.x \qquad \text{and} \qquad \lambda x^A.\lambda y^A.y \\
(5) & \lambda x^{(A\to A)\to B}.x(\lambda y^A.y) \\
(6) & \text{can't find a closed term} \\
(7) & \text{can't find a closed term}
\end{array}$$

Can we answer the general question, given a type, whether there exists a closed term for it?

For a new way to look at the problem, take the types (1)–(7) and make the following replacement of symbols: replace "$\to$" by "$\Rightarrow$" and replace "$\times$" by "$\wedge$". We obtain the following formulas:

$$\begin{array}{ll}
(1) & (A \wedge B) \Rightarrow A \\
(2) & A \Rightarrow B \Rightarrow (A \wedge B) \\
(3) & (A \Rightarrow B) \Rightarrow (B \Rightarrow C) \Rightarrow (A \Rightarrow C) \\
(4) & A \Rightarrow A \Rightarrow A \\
(5) & ((A \Rightarrow A) \Rightarrow B) \Rightarrow B \\
(6) & A \Rightarrow (A \wedge B) \\
(7) & (A \Rightarrow C) \Rightarrow C
\end{array}$$

Note that these are formulas of propositional logic, where "$\Rightarrow$" is implication, and "$\wedge$" is conjunction ("and"). What can we say about the validity of these formulas? It turns out that (1)–(5) are tautologies, whereas (6)–(7) are not. Thus, the types for which we could find a lambda term turn out to be the ones that are valid when considered as formulas in propositional logic! This is not entirely coincidental.

Let us consider, for example, how to prove $(A \wedge B) \Rightarrow A$. The proof is very short. It goes as follows: "Assume $A \wedge B$. Then, by the first part of that assumption, A holds. Thus $(A \wedge B) \Rightarrow A$." On the other hand, the lambda term of the corresponding type is $\lambda x^{A \times B}.\pi_1 x$. You can see that there is a close connection between the proof and the lambda term. Namely, if one reads $\lambda x^{A \times B}$ as "assume $A \wedge B$ (call the assumption 'x')", and if one reads $\pi_1 x$ as "by the first part of assumption x", then this lambda term can be read as a proof of the proposition $(A \wedge B) \Rightarrow A$.

This connection between the simply-typed lambda calculus and propositional logic is known as the "Curry-Howard isomorphism". Since types of the lambda calculus correspond to formulas in propositional logic, and terms correspond to proofs, the concept is also known as the "proofs-as-programs" paradigm, or the "formulas-as-types" correspondence. We will make the actual correspondence more precise in the next two sections.

Before we go any further, we must make one important point. When we are going to make precise the connection between simply-typed lambda calculus and propositional logic, we will see that the appropriate logic is *intuitionistic logic*, and not the ordinary *classical logic* that we are used to from mathematical practice. The main difference between intuitionistic and classical logic is that the former misses the principles of "proof by contradiction" and "excluded middle". The principle of proof by contradiction states that if the assumption "not A" leads to a contradiction then we have proved A. The principle of excluded middle states that either "A" or "not A" must be true.

Intuitionistic logic is also known as *constructive logic*, because all proofs in it are by construction. Thus, in intuitionistic logic, the only way to prove the existence of some object is by actually constructing the object. This is in contrast with classical logic, where we may prove the existence of an object simply by deriving a contradiction from the assumption that the object doesn't exist. The disadvantage of constructive logic is that it is generally more difficult to prove things. The advantage is that once one has a proof, the proof can be transformed into an algorithm.

6.3 Propositional intuitionistic logic

We start by introducing a system for intuitionistic logic that uses only three connectives: "$\wedge$", "$\rightarrow$", and "$\top$". *Formulas* $A, B \dots$ are built from atomic formulas $\alpha, \beta, \dots$ via the BNF

$$\text{Formulas:} \quad A, B ::= \alpha \mid A \rightarrow B \mid A \wedge B \mid \top.$$

We now need to formalize proofs. The formalized proofs will be called "derivations". The system we introduce here is known as *natural deduction*, and is due to Gentzen (1935).

In natural deduction, derivations are certain kinds of trees. In general, we will deal with derivations of a formula A from a set of assumptions $\Gamma = \{A_1, \dots, A_n\}$. Such a derivation will be written schematically as

$$\begin{array}{c} x_1{:}A_1, \dots, x_n{:}A_n \\ \vdots \\ A \end{array} \quad .$$

We simplify the bookkeeping by giving a name to each assumption, and we will use lower-case letters such as x, y, z for such names. In using the above notation for schematically writing a derivation of A from assumptions Γ, it is understood that the derivation may in fact use a given assumption more than once, or zero times. The rules for constructing derivations are as follows:

1. (Axiom)

$$(ax)\ \frac{x{:}A}{A}\,x$$

   is a derivation of A from assumption A (and possibly other assumptions that were used zero times). We have written the letter "x" next to the rule, to indicate precisely which assumption we have used here.

2. ($\wedge$-introduction) If

$$\begin{array}{c} \Gamma \\ \vdots \\ A \end{array} \quad \text{and} \quad \begin{array}{c} \Gamma \\ \vdots \\ B \end{array}$$

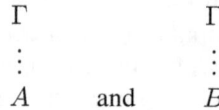

are derivations of A and B, respectively, then

$$
(\wedge\text{-}I)\ \frac{\begin{array}{c}\Gamma\\ \vdots\\ A\end{array}\qquad\begin{array}{c}\Gamma\\ \vdots\\ B\end{array}}{A \wedge B}
$$

is a derivation of $A \wedge B$. In other words, a proof of $A \wedge B$ is a proof of A and a proof of B.

3. ($\wedge$-elimination) If

$$
\begin{array}{c}\Gamma\\ \vdots\\ A \wedge B\end{array}
$$

is a derivation of $A \wedge B$, then

$$
(\wedge\text{-}E_1)\ \frac{\begin{array}{c}\Gamma\\ \vdots\\ A \wedge B\end{array}}{A} \qquad \text{and} \qquad (\wedge\text{-}E_2)\ \frac{\begin{array}{c}\Gamma\\ \vdots\\ A \wedge B\end{array}}{B}
$$

are derivations of A and B, respectively. In other words, from $A \wedge B$, we are allowed to conclude both A and B.

4. ($\top$-introduction)

$$
(\top\text{-}I)\ \frac{}{\top}
$$

is a derivation of $\top$ (possibly from some assumptions, which were not used). In other words, $\top$ is always true.

5. ($\to$-introduction) If

$$
\begin{array}{c}\Gamma, x{:}A\\ \vdots\\ B\end{array}
$$

is a derivation of B from assumptions Γ and A, then

$$
(\to\text{-}I)\ \frac{\begin{array}{c}\Gamma, [x{:}A]\\ \vdots\\ B\end{array}}{A \to B}\, x
$$

is a derivation of $A \to B$ from Γ alone. Here, the assumption $x{:}A$ is no longer an assumption of the new derivation — we say that it has been "cancelled". We indicate cancelled assumptions by enclosing them in brackets

[ ], and we indicate the place where the assumption was cancelled by writing the letter x next to the rule where it was cancelled.

6. ($\to$-elimination) If

$$\begin{array}{c} \Gamma \\ \vdots \\ A \to B \end{array} \qquad \text{and} \qquad \begin{array}{c} \Gamma \\ \vdots \\ A \end{array}$$

are derivations of $A \to B$ and A, respectively, then

$$(\to\text{-}E)\ \frac{\begin{array}{c} \Gamma \\ \vdots \\ A \to B \end{array} \qquad \begin{array}{c} \Gamma \\ \vdots \\ A \end{array}}{B}$$

is a derivation of B. In other words, from $A \to B$ and A, we are allowed to conclude B. This rule is sometimes called by its Latin name, "modus ponens".

This finishes the definition of derivations in natural deduction. Note that, with the exception of the axiom, each rule belongs to some specific logical connective, and there are introduction and elimination rules. "$\wedge$" and "$\to$" have both introduction and elimination rules, whereas "$\top$" only has an introduction rule.

In natural deduction, like in real mathematical life, assumptions can be made at any time. The challenge is to get rid of assumptions once they are made. In the end, we would like to have a derivation of a given formula that depends on as few assumptions as possible — in fact, we don't regard the formula as proven unless we can derive it from *no* assumptions. The rule ($\to$-*I*) allows us to discard temporary assumptions that we might have made during the proof.

Exercise 29. Give a derivation, in natural deduction, for each of the formulas (1)–(5) from Section 6.2.

6.4 An alternative presentation of natural deduction

The above notation for natural deduction derivations suffers from a problem of presentation: since assumptions are first written down, later cancelled dynamically, it is not easy to see when each assumption in a finished derivation was cancelled.

The following alternate presentation of natural deduction works by deriving entire *judgments*, rather than *formulas*. Rather than keeping track of assumptions as the leaves of a proof tree, we annotate each formula in a derivation with the entire set of assumptions that were used in deriving it. In practice, this makes derivations more verbose, by repeating most assumptions on each line. In theory, however, such derivations are easier to reason about.

A *judgment* is a statement of the form $x_1{:}A_1, \ldots, x_n{:}A_n \vdash B$. It states that the formula B is a consequence of the (labelled) assumptions $A_1, \ldots, A_n$. The rules of natural deduction can now be reformulated as rules for deriving judgments:

1. (Axiom)

$$(ax_x)\ \frac{}{\Gamma, x{:}A \vdash A}$$

2. ($\wedge$-introduction)

$$(\wedge\text{-}I)\ \frac{\Gamma \vdash A \qquad \Gamma \vdash B}{\Gamma \vdash A \wedge B}$$

3. ($\wedge$-elimination)

$$(\wedge\text{-}E_1)\ \frac{\Gamma \vdash A \wedge B}{\Gamma \vdash A} \qquad (\wedge\text{-}E_2)\ \frac{\Gamma \vdash A \wedge B}{\Gamma \vdash B}$$

4. ($\top$-introduction)

$$(\top\text{-}I)\ \frac{}{\Gamma \vdash \top}$$

5. ($\to$-introduction)

$$(\to\text{-}I_x)\ \frac{\Gamma, x{:}A \vdash B}{\Gamma \vdash A \to B}$$

6. ($\to$-elimination)

$$(\to\text{-}E)\ \frac{\Gamma \vdash A \to B \qquad \Gamma \vdash A}{\Gamma \vdash B}$$

6.5 The Curry-Howard Isomorphism

There is an obvious one-to-one correspondence between types of the simply-typed lambda calculus and the formulas of propositional intuitionistic logic introduced in Section 6.3 (provided that the set of basic types can be identified with the set of atomic formulas). We will identify formulas and types from now on, where it is convenient to do so.

Perhaps less obvious is the fact that derivations are in one-to-one correspondence with simply-typed lambda terms. To be precise, we will give a translation from derivations to lambda terms, and a translation from lambda terms to derivations, which are mutually inverse up to α-equivalence.

To any derivation of $x_1{:}A_1, \ldots, x_n{:}A_n \vdash B$, we will associate a lambda term M such that $x_1{:}A_1, \ldots, x_n{:}A_n \vdash M : B$ is a valid typing judgment. We define M by recursion on the definition of derivations. We prove simultaneously, by induction, that $x_1{:}A_1, \ldots, x_n{:}A_n \vdash M : B$ is indeed a valid typing judgment.

1. (Axiom) If the derivation is

$$(ax_x)\ \frac{}{\Gamma, x{:}A \vdash A},$$

then the lambda term is $M = x$. Clearly, $\Gamma, x{:}A \vdash x : A$ is a valid typing judgment by (*var*).

2. ($\wedge$-introduction) If the derivation is

$$(\wedge\text{-}I)\ \frac{\begin{array}{cc}\vdots & \vdots \\ \Gamma \vdash A & \Gamma \vdash B\end{array}}{\Gamma \vdash A \wedge B},$$

then the lambda term is $M = \langle P, Q \rangle$, where P and Q are the terms associated to the two respective subderivations. By induction hypothesis, $\Gamma \vdash P : A$ and $\Gamma \vdash Q : B$, thus $\Gamma \vdash \langle P, Q \rangle : A \times B$ by (*pair*).

3. ($\wedge$-elimination) If the derivation is

$$(\wedge\text{-}E_1)\ \frac{\begin{array}{c}\vdots \\ \Gamma \vdash A \wedge B\end{array}}{\Gamma \vdash A},$$

then we let $M = \pi_1 P$, where P is the term associated to the subderivation. By induction hypothesis, $\Gamma \vdash P : A \times B$, thus $\Gamma \vdash \pi_1 P : A$ by (π_1). The case of ($\wedge$-E_2) is entirely symmetric.

4. ($\top$-introduction) If the derivation is

$$(\top\text{-}I)\ \frac{}{\Gamma \vdash \top},$$

then let $M = *$. We have $\vdash * : 1$ by ($*$).

5. ($\to$-introduction) If the derivation is

$$(\to\text{-}I_x)\ \frac{\begin{array}{c}\vdots \\ \Gamma, x{:}A \vdash B\end{array}}{\Gamma \vdash A \to B},$$

then we let $M = \lambda x^A.P$, where P is the term associated to the subderivation. By induction hypothesis, $\Gamma, x{:}A \vdash P : B$, hence $\Gamma \vdash \lambda x^A.P : A \to B$ by (*abs*).

6. ($\to$-elimination) Finally, if the derivation is

$$(\to\text{-}E)\ \frac{\begin{array}{c}\vdots \\ \Gamma \vdash A \to B\end{array} \qquad \begin{array}{c}\vdots \\ \Gamma \vdash A\end{array}}{\Gamma \vdash B},$$

then we let $M = PQ$, where P and Q are the terms associated to the two respective subderivations. By induction hypothesis, $\Gamma \vdash P : A \to B$ and $\Gamma \vdash Q : A$, thus $\Gamma \vdash PQ : B$ by (*app*).

Conversely, given a well-typed lambda term M, with associated typing judgment $\Gamma \vdash M : A$, then we can construct a derivation of A from assumptions Γ. We define this derivation by recursion on the type derivation of $\Gamma \vdash M : A$. The details are too tedious to spell them out here; we simply go through each of the rules (*var*), (*abs*), (*app*), (*pair*), (π_1), (π_2), ($*$) and apply the corresponding rule (*ax*), ($\to$-*I*), ($\to$-*E*), ($\wedge$-*I*), ($\wedge$-E_1), ($\wedge$-E_2), ($\top$-*I*), respectively.

6.6 Reductions in the simply-typed lambda calculus

β- and η-reductions in the simply-typed lambda calculus are defined much in the same way as for the untyped lambda calculus, except that we have introduced some additional terms (such as pairs and projections), which calls for some additional reduction rules. We define the following reductions:

$$\begin{array}{llcll}
(\beta_{\to}) & (\lambda x^A.M)N & \to & M[N/x], & \\
(\eta_{\to}) & \lambda x^A.Mx & \to & M, & \text{where } x \notin FV(M), \\
(\beta_{\times,1}) & \pi_1\langle M, N\rangle & \to & M, & \\
(\beta_{\times,2}) & \pi_2\langle M, N\rangle & \to & N, & \\
(\eta_{\times}) & \langle \pi_1 M, \pi_2 M\rangle & \to & M, & \\
(\eta_1) & M & \to & *, & \text{if } M : 1.
\end{array}$$

Then single- and multi-step β- and η-reduction are defined as the usual contextual closure of the above rules, and the definitions of β- and η-equivalence also follow the usual pattern. In addition to the usual (*cong*) and (ξ) rules, we now also have congruence rules that apply to pairs and projections.

We remark that, to be perfectly precise, we should have defined reductions between typing judgments, and not between terms. This is necessary because some of the reduction rules, notably (η_1), depend on the type of the terms involved. However, this would be notationally very cumbersome, and we will blur the distinction, pretending at times that terms appear in some implicit typing context that we do not write.

An important property of the reduction is the "subject reduction" property, which states that well-typed terms reduce only to well-typed terms of the same type. This has an immediate application to programming: subject reduction guarantees that if we write a program of type "integer", then the final result of evaluating the program, if any, will indeed be an integer, and not, say, a boolean.

Theorem 6.1 (Subject Reduction). *If* $\Gamma \vdash M : A$ *and* $M \to_{\beta\eta} M'$, *then* $\Gamma \vdash M' : A$.

Proof. By induction on the derivation of $M \to_{\beta\eta} M'$, and by case distinction on the last rule used in the derivation of $\Gamma \vdash M : A$. For instance, if $M \to_{\beta\eta} M'$ by $(\beta_{\to})$, then $M = (\lambda x^B.P)Q$ and $M' = P[Q/x]$. If $\Gamma \vdash M : A$, then we must have $\Gamma, x{:}B \vdash P : A$ and $\Gamma \vdash Q : B$. It follows that $\Gamma \vdash P[Q/x] : A$; the latter statement can be proved separately (as a "substitution lemma") by induction on P and makes crucial use of the fact that x and Q have the same type.

The other cases are similar, and we leave them as an exercise. Note that, in particular, one needs to consider the (*cong*), (ξ), and other congruence rules as well. □

6.7 A word on Church-Rosser

One important theorem that does *not* hold for $\beta\eta$-reduction in the simply-typed $\lambda^{\to,\times,1}$-calculus is the Church-Rosser theorem. The culprit is the rule (η_1). For instance, if x is a variable of type $A \times 1$, then the term $M = \langle \pi_1 x, \pi_2 x\rangle$ reduces to x by $(\eta_\times)$, but also to $\langle \pi_1 x, *\rangle$ by (η_1). Both these terms are normal forms. Thus, the Church-Rosser property fails.

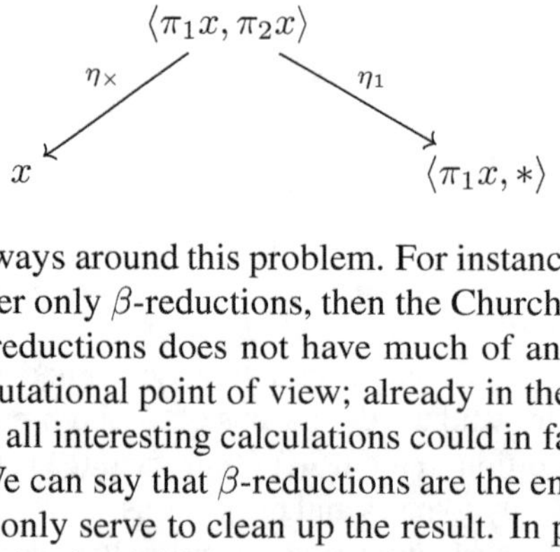

There are several ways around this problem. For instance, if we omit all the η-reductions and consider only β-reductions, then the Church-Rosser property does hold. Eliminating η-reductions does not have much of an effect on the lambda calculus from a computational point of view; already in the untyped lambda calculus, we noticed that all interesting calculations could in fact be carried out with β-reductions alone. We can say that β-reductions are the engine for computation, whereas η-reductions only serve to clean up the result. In particular, it can never happen that some η-reduction inhibits another β-reduction: if $M \to_\eta M'$, and if M' has a β-redex, then it must be the case that M already has a corresponding β-redex. Also, η-reductions always reduce the size of a term. It follows that if M is a β-normal form, then M can always be reduced to a $\beta\eta$-normal form (not necessarily unique) in a finite sequence of η-reductions.

Exercise 30. Prove the Church-Rosser theorem for β-reductions in the $\lambda^{\to,\times,1}$-calculus. Hint: use the same method that we used in the untyped case.

Another solution is to omit the type 1 and the term $*$ from the language. In this case, the Church-Rosser property holds even for $\beta\eta$-reduction.

Exercise 31. Prove the Church-Rosser theorem for $\beta\eta$-reduction in the $\lambda^{\to,\times}$-calculus, i.e., the simply-typed lambda calculus without 1 and $*$.

6.8   Reduction as proof simplification

Having made a one-to-one correspondence between simply-typed lambda terms and derivations in intuitionistic natural deduction, we may now ask what β- and η-reductions correspond to under this correspondence. It turns out that these reductions can be thought of as "proof simplification steps".

Consider for example the β-reduction $\pi_1\langle M, N\rangle \to M$. If we translate the left-hand side and the right-hand side via the Curry-Howard isomorphism (here we use the first notation for natural deduction), we get

$$
(\wedge\text{-}E_1)\dfrac{(\wedge\text{-}I)\dfrac{\begin{array}{c}\Gamma\\\vdots\\A\end{array}\quad\begin{array}{c}\Gamma\\\vdots\\B\end{array}}{A\wedge B}}{A} \quad\to\quad \begin{array}{c}\Gamma\\\vdots\\A\end{array}\ .
$$

We can see that the left derivation contains an introduction rule immediately followed by an elimination rule. This leads to an obvious simplification if we replace the left derivation by the right one.

In general, β-redexes correspond to situations where an introduction rule is immediately followed by an elimination rule, and η-redexes correspond to situations where an elimination rule is immediately followed by an introduction rule. For example, consider the η-reduction $\langle \pi_1 M, \pi_2 M\rangle \to M$. This translates to:

$$
(\wedge\text{-}I)\dfrac{(\wedge\text{-}E_1)\dfrac{\begin{array}{c}\Gamma\\\vdots\\A\wedge B\end{array}}{A}\quad(\wedge\text{-}E_2)\dfrac{\begin{array}{c}\Gamma\\\vdots\\A\wedge B\end{array}}{B}}{A\wedge B} \quad\to\quad \begin{array}{c}\Gamma\\\vdots\\A\wedge B\end{array}
$$

Again, this is an obvious simplification step, but it has a side condition: the left and right subderivation must be the same! This side condition corresponds to the fact that in the redex $\langle \pi_1 M, \pi_2 M\rangle$, the two subterms called M must be equal. It is another characteristic of η-reductions that they often carry such side conditions.

The reduction $M \to *$ translates as follows:

$$
\begin{array}{c}\Gamma\\\vdots\\\top\end{array} \quad\to\quad (\top\text{-}I)\dfrac{}{\top}
$$

In other words, any derivation of $\top$ can be replaced by the canonical such derivation.

More interesting is the case of the $(\beta_{\to})$ rule. Here, we have $(\lambda x^A.M)N \to M[N/x]$, which can be translated via the Curry-Howard Isomorphism as follows:

$$(\to\text{-}E)\ \dfrac{(\to\text{-}I)\ \dfrac{\begin{matrix}\Gamma, [x{:}A]\\ \vdots \\ B\end{matrix}}{A\to B}\,x \qquad \begin{matrix}\Gamma\\ \vdots\\ A\end{matrix}}{B} \quad\to\quad \begin{matrix}\Gamma\\ \vdots\\ \Gamma,\ A\\ \vdots\\ B\end{matrix}\ .$$

What is going on here is that we have a derivation M of B from assumptions Γ and A, and we have another derivation N of A from Γ. We can directly obtain a derivation of B from Γ by stacking the second derivation on top of the first!

Notice that this last proof "simplification" step may not actually be a simplification. Namely, if the hypothesis labelled x is used many times in the derivation M, then N will have to be copied many times in the right-hand side term. This corresponds to the fact that if x occurs several times in M, then $M[N/x]$ might be a longer and more complicated term than $(\lambda x.M)N$.

Finally, consider the $(\eta_{\to})$ rule $\lambda x^A.Mx \to M$, where $x \notin FV(M)$. This translates to derivations as follows:

$$(\to\text{-}I)\ \dfrac{(\to\text{-}E)\ \dfrac{\begin{matrix}\Gamma\\ \vdots\\ A\to B\end{matrix} \qquad (ax)\ \dfrac{[x{:}A]}{A}\,x}{B}}{A\to B}\,x \quad\to\quad \begin{matrix}\Gamma\\ \vdots\\ A\to B\end{matrix}$$

6.9 Getting mileage out of the Curry-Howard isomorphism

The Curry-Howard isomorphism makes a connection between the lambda calculus and logic. We can think of it as a connection between "programs" and "proofs". What is such a connection good for? Like any isomorphism, it allows us to switch back and forth and think in whichever system suits our intuition in a given situation. Moreover, we can save a lot of work by transferring theorems that were proved about the lambda calculus to logic, and vice versa. As an example, we will see in the next section how to add disjunctions to propositional intuitionistic logic, and then we will explore what we can learn about the lambda calculus from that.

6.10 Disjunction and sum types

To the BNF for formulas of propositional intuitionistic logic from Section 6.3, we add the following clauses:

$$\text{Formulas:}\quad A, B ::= \ldots \mid A \vee B \mid \bot.$$

Here, $A \vee B$ stands for disjunction, or "or", and $\bot$ stands for falsity, which we can also think of as zero-ary disjunction. The symbol $\bot$ is also known by the names of "bottom", "absurdity", or "contradiction". The rules for constructing derivations are extended by the following cases:

7. ($\vee$-introduction)

$$(\vee\text{-}I_1)\ \frac{\Gamma \vdash A}{\Gamma \vdash A \vee B} \qquad (\vee\text{-}I_2)\ \frac{\Gamma \vdash B}{\Gamma \vdash A \vee B}$$

In other words, if we have proven A or we have proven B, then we may conclude $A \vee B$.

8. ($\vee$-elimination)

$$(\vee\text{-}E_{x,y})\ \frac{\Gamma \vdash A \vee B \qquad \Gamma, x{:}A \vdash C \qquad \Gamma, y{:}B \vdash C}{\Gamma \vdash C}$$

This is known as the "principle of case distinction". If we know $A \vee B$, and we wish to prove some formula C, then we may proceed by cases. In the first case, we assume A holds and prove C. In the second case, we assume B holds and prove C. In either case, we prove C, which therefore holds independently.

Note that the $\vee$-elimination rule differs from all other rules we have considered so far, because it involves some arbitrary formula C that is not directly related to the principal formula $A \vee B$ being eliminated.

9. ($\bot$-elimination)

$$(\bot\text{-}E)\ \frac{\Gamma \vdash \bot}{\Gamma \vdash C},$$

for an arbitrary formula C. This rule formalizes the familiar principle "ex falsum quodlibet", which means that falsity implies anything.

There is no $\bot$-introduction rule. This is symmetric to the fact that there is no $\top$-elimination rule.

Having extended our logic with disjunctions, we can now ask what these disjunctions correspond to under the Curry-Howard isomorphism. Naturally, we need to extend the lambda calculus by as many new terms as we have new rules in the logic. It turns out that disjunctions correspond to a concept that is quite natural in programming: "sum" or "union" types.

(in_1)	$$\frac{\Gamma \vdash M : A}{\Gamma \vdash \mathrm{in}_1 M : A + B}$$
(in_2)	$$\frac{\Gamma \vdash M : B}{\Gamma \vdash \mathrm{in}_2 M : A + B}$$
$(case)$	$$\frac{\Gamma \vdash M : A + B \qquad \Gamma, x{:}A \vdash N : C \qquad \Gamma, y{:}B \vdash P : C}{\Gamma \vdash (\mathrm{case}\, M \,\mathrm{of}\, x^A \Rightarrow N \mid y^B \Rightarrow P) : C}$$
$(\Box)$	$$\frac{\Gamma \vdash M : 0}{\Gamma \vdash \Box_A M : A}$$

Table 5: Typing rules for sums

To the lambda calculus, add type constructors $A + B$ and 0.

$$\text{Simple types:} \quad A, B ::= \ldots \mid A + B \mid 0.$$

Intuitively, $A + B$ is the disjoint union of A and B, as in set theory: an element of $A + B$ is either an element of A or an element of B, together with an indication of which one is the case. In particular, if we consider an element of $A + A$, we can still tell whether it is in the left or right component, even though the two types are the same. In programming languages, this is sometimes known as a "union" or "variant" type. We call it a "sum" type here. The type 0 is simply the empty type, corresponding to the empty set in set theory.

What should the lambda terms be that go with these new types? We know from our experience with the Curry-Howard isomorphism that we have to have precisely one term constructor for each introduction or elimination rule of natural deduction. Moreover, we know that if such a rule has n subderivations, then our term constructor has to have n immediate subterms. We also know something about bound variables: Each time a hypothesis is cancelled in a natural deduction rule, there must be a binder of the corresponding variable in the lambda calculus. This information more or less uniquely determines what the lambda terms should be; the only choice that is left is what to call them!

We add four terms to the lambda calculus:

$$\begin{array}{llll}\text{Raw terms:} & M, N, P & ::= & \ldots \mid \mathrm{in}_1 M \mid \mathrm{in}_2 M \\ & & & \mid \mathrm{case}\, M \,\mathrm{of}\, x^A \Rightarrow N \mid y^B \Rightarrow P \mid \Box_A M\end{array}$$

The typing rules for these new terms are shown in Table 5. By comparing these rules to $(\vee\text{-}I_1)$, $(\vee\text{-}I_2)$, $(\vee\text{-}E)$, and $(\bot\text{-}E)$, you can see that they are precisely analogous.

But what is the meaning of these new terms? The term $\mathrm{in}_1 M$ is simply an element of the left component of $A + B$. We can think of in_1 as the injection function $A \to A + B$. Similar for in_2. The term $(\mathrm{case}\, M \,\mathrm{of}\, x^A \Rightarrow N \mid y^B \Rightarrow P)$ is a case distinction: evaluate M of type $A + B$. The answer is either an element of the left component A or of the right component B. In the first case, assign the answer to the variable x and evaluate N. In the second case, assign the answer to the variable y and evaluate P. Since both N and P are of type C, we get a final result of type C. Note that the case statement is very similar to an if-then-else; the only difference is that the two alternatives also carry a value. Indeed, the booleans can be defined as $1+1$, in which case $\mathbf{T} = \mathrm{in}_1*$, $\mathbf{F} = \mathrm{in}_2*$, and **if_then_else** $MNP = \mathrm{case}\, M \,\mathrm{of}\, x^1 \Rightarrow N \mid y^1 \Rightarrow P$, where x and y don't occur in N and P, respectively.

Finally, the term $\Box_A M$ is a simple type cast, corresponding to the unique function $\Box_A : 0 \to A$ from the empty set to any set A.

6.11 Classical logic vs. intuitionistic logic

We have mentioned before that the natural deduction calculus we have presented corresponds to intuitionistic logic, and not classical logic. But what exactly is the difference? Well, the difference is that in intuitionistic logic, we have no rule for proof by contradiction, and we do not have $A \vee \neg A$ as an axiom.

Let us adopt the following convention for negation: the formula $\neg A$ ("not A") is regarded as an abbreviation for $A \to \bot$. This way, we do not have to introduce special formulas and rules for negation; we simply use the existing rules for $\to$ and $\bot$.

In intuitionistic logic, there is no derivation of $A \vee \neg A$, for general A. Or equivalently, in the simply-typed lambda calculus, there is no closed term of type $A + (A \to 0)$. We are not yet in a position to prove this formally, but informally, the argument goes as follows: If the type A is empty, then there can be no closed term of type A (otherwise A would have that term as an element). On the other hand, if the type A is non-empty, then there can be no closed term of type $A \to 0$ (or otherwise, if we applied that term to some element of A, we would obtain an element of 0). But if we were to write a *generic* term of type $A + (A \to 0)$, then this term would have to work no matter what A is. Thus, the term would have to decide whether to use the left or right component independently of A. But for any such term, we can get a contradiction by choosing A either empty or non-empty.

Closely related is the fact that in intuitionistic logic, we do not have a principle of proof by contradiction. The "proof by contradiction" rule is the following:

$$(\mathit{contra}_x)\ \frac{\Gamma, x{:}\neg A \vdash \bot}{\Gamma \vdash A}.$$

This is *not* a rule of intuitionistic propositional logic, but we can explore what would happen if we were to add such a rule. First, we observe that the contradiction rule is very similar to the following:

$$\frac{\Gamma, x{:}A \vdash \bot}{\Gamma \vdash \neg A}.$$

However, since we defined $\neg A$ to be the same as $A \to \bot$, the latter rule is an instance of ($\to$-*I*). The contradiction rule, on the other hand, is not an instance of ($\to$-*I*).

If we admit the rule (*contra*), then $A \vee \neg A$ can be derived. The following is such a derivation:

$$
(\mathit{contra}_y)\;\dfrac{(\to\text{-}E)\;\dfrac{(ax_y)\;\dfrac{}{y{:}\neg(A\vee\neg A)\vdash\neg(A\vee\neg A)} \qquad (\vee\text{-}I_2)\;\dfrac{(\to\text{-}I_x)\;\dfrac{(\to\text{-}E)\;\dfrac{(ax_y)\;\dfrac{}{y{:}\neg(A\vee\neg A),x{:}A\vdash\neg(A\vee\neg A)} \qquad (\vee\text{-}I_1)\;\dfrac{(ax_x)\;\dfrac{}{y{:}\neg(A\vee\neg A),x{:}A\vdash A}}{y{:}\neg(A\vee\neg A),x{:}A\vdash A\vee\neg A}}{y{:}\neg(A\vee\neg A),x{:}A\vdash\bot}}{y{:}\neg(A\vee\neg A)\vdash\neg A}}{y{:}\neg(A\vee\neg A)\vdash A\vee\neg A}}{y{:}\neg(A\vee\neg A)\vdash\bot}}{\vdash A\vee\neg A}
$$

Conversely, if we added $A \vee \neg A$ as an axiom to intuitionistic logic, then this already implies the (*contra*) rule. Namely, from any derivation of $\Gamma, x{:}\neg A \vdash \bot$, we can obtain a derivation of $\Gamma \vdash A$ by using $A \vee \neg A$ as an axiom. Thus, we can *simulate* the (*contra*) rule, in the presence of $A \vee \neg A$.

$$
(\vee\text{-}E_{x,y})\;\dfrac{\dfrac{(\textit{excluded middle})}{\Gamma\vdash A\vee\neg A} \qquad (\bot\text{-}E)\;\dfrac{\Gamma,x{:}\neg A\vdash\bot}{\Gamma,x{:}\neg A\vdash A} \qquad (ax_y)\;\dfrac{}{\Gamma,y{:}A\vdash A}}{\Gamma\vdash A}
$$

In this sense, we can say that the rule (*contra*) and the axiom $A \vee \neg A$ are equivalent, in the presence of the other axioms and rules of intuitionistic logic.

It turns out that the system of intuitionistic logic plus (*contra*) is equivalent to classical logic as we know it. It is in this sense that we can say that intuitionistic logic is "classical logic without proofs by contradiction".

Exercise 32. The formula $((A \to B) \to A) \to A$ is called "Peirce's law". It is valid in classical logic, but not in intuitionistic logic. Give a proof of Peirce's law in natural deduction, using the rule (*contra*).

Conversely, Peirce's law, when added to intuitionistic logic for all A and B, implies (*contra*). Here is the proof. Recall that $\neg A$ is an abbreviation for $A \to \bot$.

$$
(\to\text{-}E)\;\dfrac{\dfrac{(\textit{Peirce's law for } B=\bot)}{\Gamma\vdash((A\to\bot)\to A)\to A} \qquad (\to\text{-}I_x)\;\dfrac{(\bot\text{-}E)\;\dfrac{\Gamma,x{:}A\to\bot\vdash\bot}{\Gamma,x{:}A\to\bot\vdash A}}{\Gamma\vdash(A\to\bot)\to A}}{\Gamma\vdash A}
$$

We summarize the results of this section in terms of a slogan:

$$
\begin{array}{rl}
 & \text{intuitionistic logic + } (\textit{contra}) \\
= & \text{intuitionistic logic + ``}A\vee\neg A\text{''} \\
= & \text{intuitionistic logic + Peirce's law} \\
= & \text{classical logic.}
\end{array}
$$

The proof theory of intuitionistic logic is a very interesting subject in its own right, and an entire course could be taught just on that subject.

6.12 Classical logic and the Curry-Howard isomorphism

To extend the Curry-Howard isomorphism to classical logic, according to the observations of the previous section, it is sufficient to add to the lambda calculus a term representing Peirce's law. All we have to do is to add a term $\mathfrak{C} : ((A \to B) \to A) \to A$, for all types A and B.

Such a term is known as *Felleisen's* $\mathfrak{C}$, and it has a specific interpretation in terms of programming languages. It can be understood as a control operator (similar to "goto", "break", or exception handling in some procedural programming languages).

Specifically, Felleisen's interpretation requires a term of the form

$$M = \mathfrak{C}(\lambda k^{A \to B}.N) : A$$

to be evaluated as follows. To evaluate M, first evaluate N. Note that both M and N have type A. If N returns a result, then this immediately becomes the result of M as well. On the other hand, if during the evaluation of N, the function k is ever called with some argument $x : A$, then the further evaluation of N is aborted, and x immediately becomes the result of M.

In other words, the final result of M can be calculated anywhere inside N, no matter how deeply nested, by passing it to k as an argument. The function k is known as a *continuation*.

There is a lot more to programming with continuations than can be explained in these lecture notes. For an interesting application of continuations to compiling, see e.g. [9] from the bibliography (Section 15). The above explanation of what it means to "evaluate" the term M glosses over several details. In particular, we have not given a reduction rule for $\mathfrak{C}$ in the style of β-reduction. To do so is rather complicated and is beyond the scope of these notes.

7 Weak and strong normalization

7.1 Definitions

As we have seen, computing with lambda terms means reducing lambda terms to normal form. By the Church-Rosser theorem, such a normal form is guaranteed to be unique if it exists. But so far, we have paid little attention to the question whether normal forms exist for a given term, and if so, how we need to reduce the term to find a normal form.

Definition. Given a notion of term and a reduction relation, we say that a term M is *weakly normalizing* if there exists a finite sequence of reductions $M \to M_1 \to \ldots \to M_n$ such that M_n is a normal form. We say that M is *strongly normalizing* if there does not exist an infinite sequence of reductions starting from M, or in other words, if *every* sequence of reductions starting from M is finite.

Recall the following consequence of the Church-Rosser theorem, which we stated as Corollary 4.2: If M has a normal form N, then $M \twoheadrightarrow N$. It follows that a term M is weakly normalizing if and only if it has a normal form. This does not imply that every possible way of reducing M leads to a normal form. A term is strongly normalizing if and only if every way of reducing it leads to a normal form in finitely many steps.

Consider for example the following terms in the untyped lambda calculus:

1. The term $\Omega = (\lambda x.xx)(\lambda x.xx)$ is neither weakly nor strongly normalizing. It does not have a normal form.

2. The term $(\lambda x.y)\Omega$ is weakly normalizing, but not strongly normalizing. It reduces to the normal form y, but it also has an infinite reduction sequence.

3. The term $(\lambda x.y)((\lambda x.x)(\lambda x.x))$ is strongly normalizing. While there are several different ways to reduce this term, they all lead to a normal form in finitely many steps.

4. The term $\lambda x.x$ is strongly normalizing, since it has no reductions, much less an infinite reduction sequence. More generally, every normal form is strongly normalizing.

We see immediately that strongly normalizing implies weakly normalizing. However, as the above examples show, the converse is not true.

7.2 Weak and strong normalization in typed lambda calculus

We found that the term $\Omega = (\lambda x.xx)(\lambda x.xx)$ is not weakly or strongly normalizing. On the other hand, we also know that this term is not typable in the simply-typed lambda calculus. This is not a coincidence, as the following theorem shows.

Theorem 7.1 (Weak normalization theorem). *In the simply-typed lambda calculus, all terms are weakly normalizing.*

Theorem 7.2 (Strong normalization theorem). *In the simply-typed lambda calculus, all terms are strongly normalizing.*

Clearly, the strong normalization theorem implies the weak normalization theorem. However, the weak normalization theorem is much easier to prove, which is the reason we proved both these theorems in class. In particular, the proof of the weak normalization theorem gives an explicit measure of the complexity of a term, in terms of the number of redexes of a certain degree in the term. There is no corresponding complexity measure in the proof of the strong normalization theorem.

Please refer to Chapters 4 and 6 of "Proofs and Types" by Girard, Lafont, and Taylor [2] for the proofs of Theorems 7.1 and 7.2, respectively.

8 Polymorphism

The polymorphic lambda calculus, also known as "System F", is obtained extending the Curry-Howard isomorphism to the quantifier $\forall$. For example, consider the identity function $\lambda x^A.x$. This function has type $A \to A$. Another identity function is $\lambda x^B.x$ of type $B \to B$, and so forth for every type. We can thus think of the identity function as a family of functions, one for each type. In the polymorphic lambda calculus, there is a dedicated syntax for such families, and we write $\Lambda\alpha.\lambda x^\alpha.x$ of type $\forall\alpha.\alpha \to \alpha$.

System F was independently discovered by Jean-Yves Girard and John Reynolds in the early 1970s.

8.1 Syntax of System F

The primary difference between System F and simply-typed lambda calculus is that System F has a new kind of function that takes a *type*, rather than a *term*, as its argument. We can also think of such a function as a family of terms that is indexed by a type.

Let α, β, γ range over a countable set of *type variables*. The types of System F are given by the grammar

$$\text{Types:} \quad A, B ::= \alpha \mid A \to B \mid \forall\alpha.A$$

A type of the form $A \to B$ is called a *function type*, and a type of the form $\forall\alpha.A$ is called a *universal type*. The type variable α is bound in $\forall\alpha.A$, and we identify types up to renaming of bound variables; thus, $\forall\alpha.\alpha \to \alpha$ and $\forall\beta.\beta \to \beta$ are the same type. We write $FTV(A)$ for the set of free type variables of a type A, defined inductively by:

- $FTV(\alpha) = \{\alpha\}$,
- $FTV(A \to B) = FTV(A) \cup FTV(B)$,
- $FTV(\forall\alpha.A) = FTV(A) \setminus \{\alpha\}$.

We also write $A[B/\alpha]$ for the result of replacing all free occurrences of α by B in A. Just like the substitution of terms (see Section 2.3), this type substitution must be *capture-free*, i.e., special care must be taken to rename any bound variables of A so that their names are different from the free variables of B.

The terms of System F are:

$$\text{Terms:} \quad M, N ::= x \mid MN \mid \lambda x^A.M \mid MA \mid \Lambda\alpha.M$$

Of these, variables x, applications MN, and lambda abstractions $\lambda x^A.M$ are exactly as for the simply-typed lambda calculus. The new terms are *type application*

$$
\begin{array}{ll}
(\textit{var}) & \dfrac{}{\Gamma, x{:}A \vdash x : A} \\[2ex]
(\textit{app}) & \dfrac{\Gamma \vdash M : A \to B \qquad \Gamma \vdash N : A}{\Gamma \vdash MN : B} \\[2ex]
(\textit{abs}) & \dfrac{\Gamma, x{:}A \vdash M : B}{\Gamma \vdash \lambda x^A.M : A \to B} \\[2ex]
(\textit{typeapp}) & \dfrac{\Gamma \vdash M : \forall\alpha.A}{\Gamma \vdash MB : A[B/\alpha]} \\[2ex]
(\textit{typeabs}) & \dfrac{\Gamma \vdash M : A \qquad \alpha \notin \mathit{FTV}(\Gamma)}{\Gamma \vdash \Lambda\alpha.M : \forall\alpha.A}
\end{array}
$$

Table 6: Typing rules for System F

MA, which is the application of a type function M to a type A, and *type abstraction* $\Lambda\alpha.M$, which denotes the type function that maps a type α to a term M. The typing rules for System F are shown in Table 6.

We also write $FTV(M)$ for the set of free type variables in the term M. We need a final notion of substitution: if M is a term, B a type, and α a type variable, we write $M[B/\alpha]$ for the capture-free substitution of B for α in M.

8.2 Reduction rules

In System F, there are two rules for β-reduction. The first one is the familiar rule for the application of a function to a term. The second one is an analogous rule for the application of a type function to a type.

$$
\begin{array}{llll}
(\beta_\to) & (\lambda x^A.M)N & \to & M[N/x], \\
(\beta_\forall) & (\Lambda\alpha.M)A & \to & M[A/\alpha],
\end{array}
$$

Similarly, there are two rules for η-reduction.

$$
\begin{array}{lllll}
(\eta_\to) & \lambda x^A.Mx & \to & M, & \text{if } x \notin FV(M), \\
(\eta_\forall) & \Lambda\alpha.M\alpha & \to & M, & \text{if } \alpha \notin FTV(M).
\end{array}
$$

The congruence and ξ-rules are as expected:

$$
\frac{M \to M'}{MN \to M'N} \qquad \frac{N \to N'}{MN \to MN'} \qquad \frac{M \to M'}{\lambda x^A M \to \lambda x^A M'}
$$

$$
\frac{M \to M'}{MA \to M'A} \qquad \frac{M \to M'}{\Lambda\alpha M \to \Lambda\alpha M'}
$$

8.3 Examples

Just as in the untyped lambda calculus, many interesting data types and operations can be encoded in System F.

8.3.1 Booleans

Define the System F type **bool**, and terms $\mathbf{T}, \mathbf{F} : \mathbf{bool}$, as follows:

$$\begin{array}{lcl} \mathbf{bool} & = & \forall\alpha.\alpha \to \alpha \to \alpha, \\ \mathbf{T} & = & \Lambda\alpha.\lambda x^{\alpha}.\lambda y^{\alpha}.x, \\ \mathbf{F} & = & \Lambda\alpha.\lambda x^{\alpha}.\lambda y^{\alpha}.y. \end{array}$$

It is easy to see from the typing rules that $\vdash \mathbf{T} : \mathbf{bool}$ and $\vdash \mathbf{F} : \mathbf{bool}$ are valid typing judgments. We can define an if-then-else operation

$$\begin{array}{l} \mathbf{if\_then\_else} : \forall\beta.\,\mathbf{bool} \to \beta \to \beta \to \beta, \\ \mathbf{if\_then\_else} = \Lambda\beta.\lambda z^{\mathbf{bool}}.z\beta. \end{array}$$

It is then easy to see that, for any type B and any pair of terms $M, N : B$, we have

$$\begin{array}{lcl} \mathbf{if\_then\_else}\; B\,\mathbf{T}\,MN & \twoheadrightarrow_{\beta} & M, \\ \mathbf{if\_then\_else}\; B\,\mathbf{F}\,MN & \twoheadrightarrow_{\beta} & N. \end{array}$$

Once we have if-then-else, it is easy to define other boolean operations, for example

$$\begin{array}{lcl} \mathbf{and} & = & \lambda a^{\mathbf{bool}}.\lambda b^{\mathbf{bool}}.\,\mathbf{if\_then\_else}\;\mathbf{bool}\;a\,b\,\mathbf{F}, \\ \mathbf{or} & = & \lambda a^{\mathbf{bool}}.\lambda b^{\mathbf{bool}}.\,\mathbf{if\_then\_else}\;\mathbf{bool}\;a\,\mathbf{T}\,b, \\ \mathbf{not} & = & \lambda a^{\mathbf{bool}}.\,\mathbf{if\_then\_else}\;\mathbf{bool}\;a\,\mathbf{F}\,\mathbf{T}. \end{array}$$

Later, in Proposition 8.8, we will show that up to $\beta\eta$ equality, $\mathbf{T}$ and and $\mathbf{F}$ are the *only* closed terms of type **bool**. This, together with the if-then-else operation, justifies calling this the type of booleans.

Note that the above encodings of the booleans and their if-then-else operation in System F is exactly the same as the corresponding encodings in the untyped lambda calculus from Section 3.1, provided that one erases all the types and type abstractions. However, there is an important difference: in the untyped lambda calculus, the booleans were just two terms among many, and there was no guarantee that the argument of a boolean function (such as **and** and **or**) was actually a boolean. In System F, the typing guarantees that all closed boolean terms eventually reduce to either $\mathbf{T}$ or $\mathbf{F}$.

8.3.2 Natural numbers

We can also define a type of Church numerals in System F. We define:

$$\begin{array}{lcl} \mathbf{nat} & = & \forall\alpha.(\alpha \to \alpha) \to \alpha \to \alpha, \\ \overline{0} & = & \Lambda\alpha.\lambda f^{\alpha\to\alpha}.\lambda x^{\alpha}.x, \\ \overline{1} & = & \Lambda\alpha.\lambda f^{\alpha\to\alpha}.\lambda x^{\alpha}.fx, \\ \overline{2} & = & \Lambda\alpha.\lambda f^{\alpha\to\alpha}.\lambda x^{\alpha}.f(fx), \\ \overline{3} & = & \Lambda\alpha.\lambda f^{\alpha\to\alpha}.\lambda x^{\alpha}.f(f(fx)), \\ \ldots & & \end{array}$$

It is then easy to define simple functions, such as successor, addition, and multiplication:

$$\begin{array}{rcl}\textbf{succ} & = & \lambda n^{\textbf{nat}}.\Lambda\alpha.\lambda f^{\alpha\to\alpha}.\lambda x^{\alpha}.f(n\alpha fx),\\ \textbf{add} & = & \lambda n^{\textbf{nat}}.\lambda m^{\textbf{nat}}.\Lambda\alpha.\lambda f^{\alpha\to\alpha}.\lambda x^{\alpha}.n\alpha f(m\alpha fx),\\ \textbf{mult} & = & \lambda n^{\textbf{nat}}.\lambda m^{\textbf{nat}}.\Lambda\alpha.\lambda f^{\alpha\to\alpha}.n\alpha(m\alpha f).\end{array}$$

Just as for the booleans, these encodings of the Church numerals and functions are exactly the same as those of the untyped lambda calculus from Section 3.2, if one erases all the types and type abstractions. We will show in Proposition 8.9 below that the Church numerals are, up to $\beta\eta$-equivalence, the only closed terms of type **nat**.

8.3.3 Pairs

You will have noticed that we didn't include a cartesian product type $A \times B$ in the definition of System F. This is because such a type is definable. Specifically, let

$$\begin{array}{rcl}A \times B & = & \forall\alpha.(A \to B \to \alpha) \to \alpha,\\ \langle M, N\rangle & = & \Lambda\alpha.\lambda f^{A\to B\to\alpha}.fMN.\end{array}$$

Note that when $M : A$ and $N : B$, then $\langle M, N\rangle : A \times B$. Moreover, for any pair of types A, B, we have projection functions $\pi_1 AB : A \times B \to A$ and $\pi_2 AB : A \times B \to B$, defined by

$$\begin{array}{rcl}\pi_1 & = & \Lambda\alpha.\Lambda\beta.\lambda p^{\alpha\times\beta}.p\alpha(\lambda x^{\alpha}.\lambda y^{\beta}.x),\\ \pi_2 & = & \Lambda\alpha.\Lambda\beta.\lambda p^{\alpha\times\beta}.p\beta(\lambda x^{\alpha}.\lambda y^{\beta}.y).\end{array}$$

This satisfies the usual laws

$$\begin{array}{rcl}\pi_1 AB\langle M, N\rangle & \twoheadrightarrow_\beta & M,\\ \pi_2 AB\langle M, N\rangle & \twoheadrightarrow_\beta & N.\end{array}$$

Once again, these encodings of pairs and projections are exactly the same as those we used in the untyped lambda calculus, when one erases all the type-related parts of the terms. You will show in Exercise 36 that every closed term of type $A \times B$ is $\beta\eta$-equivalent to a term of the form $\langle M, N\rangle$.

Remark 8.1. It is also worth noting that the corresponding η-laws, such as

$$\langle \pi_1 AB\, M, \pi_2 AB\, M\rangle = M,$$

are *not* derivable in System F. These laws hold whenever M is a closed term, but not necessarily when M contains free variables.

Exercise 33. Find suitable encodings in System F of the types 1, $A + B$, and 0, along with the corresponding terms $*$, in_1, in_2, $\text{case}\ M\ \text{of}\ x^A \Rightarrow N \mid y^B \Rightarrow P$, and $\Box_A M$.

8.4 Church-Rosser property and strong normalization

Theorem 8.2 (Church-Rosser). *System F satisfies the Church-Rosser property, both for β-reduction and for $\beta\eta$-reduction.*

Theorem 8.3 (Strong normalization). *In System F, all terms are strongly normalizing.*

The proof of the Church-Rosser property is similar to that of the simply-typed lambda calculus, and is left as an exercise. The proof of strong normalization is much more complex; it can be found in Chapter 14 of "Proofs and Types" [2].

8.5 The Curry-Howard isomorphism

From the point of view of the Curry-Howard isomorphism, $\forall\alpha.A$ is the universally quantified logical statement "for all α, A is true". Here α ranges over atomic propositions. For example, the formula $\forall\alpha.\forall\beta.\alpha \to (\beta \to \alpha)$ expresses the valid fact that the implication $\alpha \to (\beta \to \alpha)$ is true for all propositions α and β. Since this quantifier ranges over *propositions*, it is called a *second-order quantifier*, and the corresponding logic is *second-order propositional logic*.

Under the Curry-Howard isomorphism, the typing rules for System F become the following logical rules:

- (Axiom)

$$(ax_x)\ \frac{}{\Gamma, x{:}A \vdash A}$$

- ($\to$-introduction)

$$(\to\text{-}I_x)\ \frac{\Gamma, x{:}A \vdash B}{\Gamma \vdash A \to B}$$

- ($\to$-elimination)

$$(\to\text{-}E)\ \frac{\Gamma \vdash A \to B \qquad \Gamma \vdash A}{\Gamma \vdash B}$$

- ($\forall$-introduction)

$$(\forall\text{-}I)\ \frac{\Gamma \vdash A \qquad \alpha \notin FTV(\Gamma)}{\Gamma \vdash \forall\alpha.A}$$

- ($\forall$-elimination)

$$(\forall\text{-}E)\ \frac{\Gamma \vdash \forall\alpha.A}{\Gamma \vdash A[B/\alpha]}$$

The first three of these rules are familiar from propositional logic.

The $\forall$-introduction rule is also known as *universal generalization*. It corresponds to a well-known logical reasoning principle: If a statement A has been proven for some *arbitrary* α, then it follows that it holds for *all* α. The requirement that α is "arbitrary" has been formalized in the logic by requiring that α does not appear in any of the hypotheses that were used to derive A, or in other words, that α is not among the free type variables of Γ.

The $\forall$-elimination rule is also known as *universal specialization*. It is the simple principle that if some statement is true for all propositions α, then the same statement is true for any particular proposition B. Note that, unlike the $\forall$-introduction rule, this rule does not require a side condition.

Finally, we note that the side condition in the $\forall$-introduction rule is of course the same as that of the typing rule (*typeabs*) of Table 6. From the point of view of logic, the side condition is justified because it asserts that α is "arbitrary", i.e., no assumptions have been made about it. From a lambda calculus view, the side condition also makes sense: otherwise, the term $\lambda x^{\alpha}.\Lambda\alpha.x$ would be well-typed of type $\alpha \to \forall\alpha.\alpha$, which clearly does not make any sense: there is no way that an element x of some fixed type α could suddenly become an element of an arbitrary type.

8.6 Supplying the missing logical connectives

It turns out that a logic with only implication $\to$ and a second-order universal quantifier $\forall$ is sufficient for expressing all the other usual logical connectives, for example:

$$\begin{aligned}
A \wedge B &\iff \forall\alpha.(A \to B \to \alpha) \to \alpha, && (1)\\
A \vee B &\iff \forall\alpha.(A \to \alpha) \to (B \to \alpha) \to \alpha, && (2)\\
\neg A &\iff \forall\alpha.A \to \alpha, && (3)\\
\top &\iff \forall\alpha.\alpha \to \alpha, && (4)\\
\bot &\iff \forall\alpha.\alpha, && (5)\\
\exists\beta.A &\iff \forall\alpha.(\forall\beta.(A \to \alpha)) \to \alpha. && (6)
\end{aligned}$$

Exercise 34. Using informal intuitionistic reasoning, prove that the left-hand side is logically equivalent to the right-hand side for each of (1)–(6).

Remark 8.4. The definitions (1)–(6) are somewhat reminiscent of De Morgan's laws and double negations. Indeed, if we replace the type variable α by the constant $\mathbf{F}$ in (1), the right-hand side becomes $(A \to B \to \mathbf{F}) \to \mathbf{F}$, which is intuitionistically equivalent to $\neg\neg(A \wedge B)$. Similarly, the right-hand side of (2) becomes $(A \to \mathbf{F}) \to (B \to \mathbf{F}) \to \mathbf{F}$, which is intuitionistically equivalent to $\neg(\neg A \wedge \neg B)$, and similarly for the remaining connectives. However, the versions of (1), (2), and (6) using $\mathbf{F}$ are only *classically*, but not *intuitionistically* equivalent to their respective left-hand sides. On the other hand, it is remarkable that by the use of $\forall\alpha$, each right-hand side is *intuitionistically* equivalent to the left-hand sides.

Remark 8.5. Note the resemblance between (1) and the definition of $A \times B$ given in Section 8.3.3. Naturally, this is not a coincidence, as logical conjunction $A \wedge B$ should correspond to cartesian product $A \times B$ under the Curry-Howard correspondence. Indeed, by applying the same principle to the other logical connectives, one arrives at a good hint for Exercise 33.

Exercise 35. Extend System F with an existential quantifier $\exists\beta.A$, not by using (6), but by adding a new type with explicit introduction and elimination rules to the language. Justify the resulting rules by comparing them with the usual rules of mathematical reasoning for "there exists". Can you explain the meaning of the type $\exists\beta.A$ from a programming language or lambda calculus point of view?

8.7 Normal forms and long normal forms

Recall that a β-normal form of System F is, by definition, a term that contains no β-redex, i.e., no subterm of the form $(\lambda x^A.M)N$ or $(\Lambda\alpha.M)A$. The following proposition gives another useful way to characterize the β-normal forms.

Proposition 8.6 (Normal forms). *A term of System F is a β-normal form if and only if it is of the form*

$$\Lambda a_1.\Lambda a_2\ldots\Lambda a_n.zQ_1Q_2\ldots Q_k, \tag{7}$$

where:

- $n \geqslant 0$ and $k \geqslant 0$;
- Each Λa_i is either a lambda abstraction $\lambda x_i^{A_i}$ or a type abstraction $\Lambda\alpha_i$;
- Each Q_j is either a term M_j or a type B_j; and
- Each Q_j, when it is a term, is recursively in normal form.

Proof. First, it is clear that every term of the form (7) is in normal form: the term cannot itself be a redex, and the only place where a redex could occur is inside one of the Q_j, but these are assumed to be normal.

For the converse, consider a term M in β-normal form. We show that M is of the form (7) by induction on M.

- If $M = z$ is a variable, then it is of the form (7) with $n = 0$ and $k = 0$.
- If $M = NP$ is normal, then N is normal, so by induction hypothesis, N is of the form (7). But since NP is normal, N cannot be a lambda abstraction, so we must have $n = 0$. It follows that $NP = zQ_1Q_2\ldots Q_kP$ is itself of the form (7).
- If $M = \lambda x^A.N$ is normal, then N is normal, so by induction hypothesis, N is of the form (7). It follows immediately that $\lambda x^A.N$ is also of the form (7).

- The case for $M = NA$ is like the case for $M = NP$.
- The case for $M = \Lambda\alpha.N$ is like the case for $M = \lambda x^A.N$. □

Definition. In a term of the form (7), the variable z is called the *head variable* of the term.

Of course, by the Church-Rosser property together with strong normalization, it follows that every term of System F is β-equivalent to a unique β-normal form, which must then be of the form (7). On the other hand, the normal forms (7) are not unique up to η-conversion; for example, $\lambda x^{A\to B}.x$ and $\lambda x^{A\to B}.\lambda y^A.xy$ are η-equivalent terms and are both of the form (7). In order to achieve uniqueness up to $\beta\eta$-conversion, we introduce the notion of a *long normal form.*

Definition. A term of System F is a *long normal form* if

- it is of the form (7);
- the body $zQ_1 \dots Q_k$ is of atomic type (i.e., its type is a type variable); and
- each Q_j, when it is a term, is recursively in long normal form.

Proposition 8.7. *Every term of System F is $\beta\eta$-equivalent to a unique long normal form.*

Proof. By strong normalization and the Church-Rosser property of β-reduction, we already know that every term is β-equivalent to a unique β-normal form. It therefore suffices to show that every β-normal form is η-equivalent to a unique long normal form.

We first show that every β-normal form is η-equivalent to some long normal form. We prove this by induction. Indeed, consider a β-normal form of the form (7). By induction hypothesis, each of $Q_1, \dots, Q_k$ can be η-converted to long normal form. Now we proceed by induction on the type A of $zQ_1 \dots Q_k$. If $A = \alpha$ is atomic, then the normal form is already long, and there is nothing to show. If $A = B \to C$, then we can η-expand (7) to

$$\Lambda a_1.\Lambda a_2 \dots \Lambda a_n.\lambda w^B.zQ_1Q_2 \dots Q_k w$$

and proceed by the inner induction hypothesis. If $A = \forall\alpha.B$, then we can η-expand (7) to

$$\Lambda a_1.\Lambda a_2 \dots \Lambda a_n.\Lambda\alpha.zQ_1Q_2 \dots Q_k\alpha$$

and proceed by the inner induction hypothesis.

For uniqueness, we must show that no two different long normal forms can be $\beta\eta$-equivalent to each other. We leave this as an exercise. □

8.8 The structure of closed normal forms

It is a remarkable fact that if M is in long normal form, then a lot of the structure of M is completely determined by its type. Specifically: if the type of M is atomic, then M must start with a head variable. If the type of M is of the form $B \to C$, then M must be, up to α-equivalence, of the form $\lambda x^B.N$, where N is a long normal form of type C. And if the type of M is of the form $\forall\alpha.C$, then M must be, up to α-equivalence, of the form $\Lambda\alpha.N$, where N is a long normal form of type C.

So for example, consider the type

$$A = B_1 \to B_2 \to \forall\alpha_3.B_4 \to \forall\alpha_5.\beta.$$

We say that this type have five *prefixes*, where each prefix is of the form "$B_i \to$" or "$\forall\alpha_i.$". Therefore, every long normal form of type A must also start with five prefixes; specifically, it must start with

$$\lambda x_1^{B_1}.\lambda x_2^{B_2}.\Lambda\alpha_3.\lambda x_4^{B_4}.\Lambda\alpha_5.\ldots$$

The next part of the long normal form is a choice of head variable. If the term is closed, the head variable must be one of the x_1, x_2, or x_4. Once the head variable has been chosen, then *its* type determines how many arguments $Q_1, \ldots, Q_k$ the head variable must be applied to, and the types of these arguments. The structure of each of $Q_1, \ldots, Q_k$ is then recursively determined by its type, with its own choice of head variable, which then recursively determines its subterms, and so on.

In other words, the degree of freedom in a long normal form is a choice of head variable at each level. This choice of head variables completely determines the long normal form.

Perhaps the preceding discussion can be made more comprehensible by means of some concrete examples. The examples take the form of the following propositions and their proofs.

Proposition 8.8. *Every closed term of type* **bool** *is $\beta\eta$-equivalent to either* **T** *or* **F**.

Proof. Let M be a closed term of type **bool**. By Proposition 8.7, we may assume that M is a long normal form. Since **bool** $= \forall\alpha.\alpha \to \alpha \to \alpha$, every long normal form of this type must start, up to α-equivalence, with

$$\Lambda\alpha.\lambda x^\alpha.\lambda y^\alpha.\ldots$$

This must be followed by a head variable, which, since M is closed, can only be x or y. Since both x and y have atomic type, neither of them can be applied to further arguments, and therefore, the only two possible long normal forms are:

$$\Lambda\alpha.\lambda x^\alpha.\lambda y^\alpha.x$$
$$\Lambda\alpha.\lambda x^\alpha.\lambda y^\alpha.y,$$

which are **T** and **F**, respectively. □

Proposition 8.9. *Every closed term of type* **nat** *is* $\beta\eta$*-equivalent to a Church numeral* $\overline{n}$*, for some* $n \in \mathbb{N}$.

Proof. Let M be a closed term of type **nat**. By Proposition 8.7, we may assume that M is a long normal form. Since $\mathbf{nat} = \forall\alpha.(\alpha \to \alpha) \to \alpha \to \alpha$, every long normal form of this type must start, up to α-equivalence, with

$$\Lambda\alpha.\lambda f^{\alpha\to\alpha}.\lambda x^{\alpha}.\ldots$$

This must be followed by a head variable, which, since M is closed, can only be x or f. If the head variable is x, then it takes no argument, and we have

$$M = \Lambda\alpha.\lambda f^{\alpha\to\alpha}.\lambda x^{\alpha}.x$$

If the head variable is f, then it takes exactly one argument, so M is of the form

$$M = \Lambda\alpha.\lambda f^{\alpha\to\alpha}.\lambda x^{\alpha}.fQ_1.$$

Because Q_1 has type α, its own long normal form has no prefix; therefore Q_1 must start with a head variable, which must again be x or f. If $Q_1 = x$, we have

$$M = \Lambda\alpha.\lambda f^{\alpha\to\alpha}.\lambda x^{\alpha}.fx.$$

If Q_1 has head variable f, then we have $Q_1 = fQ_2$, and proceeding in this manner, we find that M has to be of the form

$$M = \Lambda\alpha.\lambda f^{\alpha\to\alpha}.\lambda x^{\alpha}.f(f(\ldots(fx)\ldots)),$$

i.e., a Church numeral. □

Exercise 36. Prove that every closed term of type $A \times B$ is $\beta\eta$-equivalent to a term of the form $\langle M, N\rangle$, where $M : A$ and $N : B$.

8.9 Application: representation of arbitrary data in System F

Let us consider the definition of a long normal form one more time. By definition, every long normal form is of the form

$$\Lambda a_1.\Lambda a_2 \ldots \Lambda a_n.zQ_1Q_2\ldots Q_k, \tag{8}$$

where $zQ_1Q_2\ldots Q_k$ has atomic type and $Q_1,\ldots,Q_k$ are, recursively, long normal forms. Instead of writing the long normal form on a single line as in (8), let us write it in tree form instead:

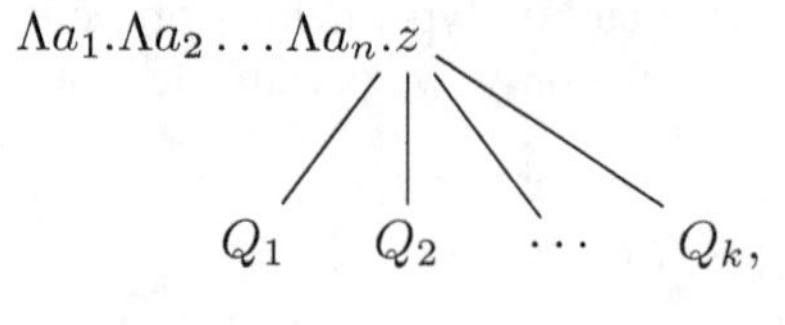

where the long normal forms $Q_1, \ldots, Q_k$ are recursively also written as trees. For example, with this notation, the Church numeral $\overline{2}$ becomes

$$\begin{array}{rl} \Lambda\alpha.\lambda f^{\alpha\to\alpha}.\lambda x^{\alpha}. & f \\ & | \\ & f \\ & | \\ & x, \end{array} \tag{9}$$

and the pair $\langle M, N \rangle$ becomes

$$\begin{array}{rccc} \Lambda\alpha.\lambda f^{A\to B\to\alpha}. & & f & \\ & / & & \backslash \\ & M & & N. \end{array}$$

We can use this very idea to encode (almost) arbitrary data structures. For example, suppose that the data structure we wish to encode is a binary tree whose leaves are labelled by natural numbers. Let's call such a thing a *leaf-labelled binary tree*. Here is an example:

$$\begin{array}{ccccc} & \bullet & & & \\ / & & \backslash & & \\ 5 & & \bullet & & \\ & / & & \backslash & \\ & 8 & & 7. & \end{array} \tag{10}$$

In general, every leaf-labelled binary tree is either a *leaf*, which is labelled by a natural number, or else a *branch* that has exactly two *children* (a left one and a right one), each of which is a leaf-labelled binary tree. Written as a BNF, we have the following grammar for leaf-labelled binary trees:

$$\text{Tree:} \quad T, S ::= \mathbf{leaf}\,(n) \;\big|\; \mathbf{branch}\,(T, S).$$

When translating this as a System F type, we think along the lines of long normal forms. We need a type variable α to represent leaf-labelled binary trees. We need two head variables whose type ends in α: The first head variable, let's call it ℓ, represents a leaf, and takes a single argument that is a natural number. Thus $\ell : \mathbf{nat} \to \alpha$. The second head variable, let's call it b, represents a branch, and takes two arguments that are leaf-labelled binary trees. Thus $b : \alpha \to \alpha \to \alpha$. We end up with the following System F type:

$$\mathbf{tree} = \forall\alpha.(\mathbf{nat} \to \alpha) \to (\alpha \to \alpha \to \alpha) \to \alpha.$$

A typical long normal form of this type is:

$$\Lambda\alpha.\lambda\ell^{\,\mathbf{nat}\,\to\alpha}.\lambda b^{\alpha\to\alpha\to\alpha}.b$$

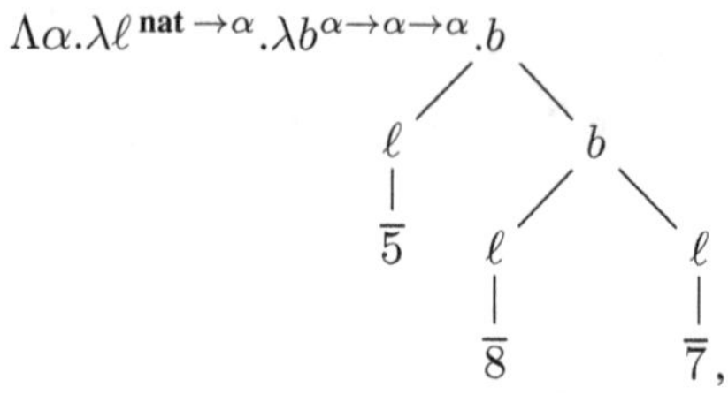

where $\overline{5}$, $\overline{7}$, and $\overline{8}$ denote Church numerals as in (9), here not expanded into long normal form for brevity. Notice how closely this long normal form follows (10). Here is the same term written on a single line:

$$\Lambda\alpha.\lambda\ell^{\,\mathbf{nat}\,\to\alpha}.\lambda b^{\alpha\to\alpha\to\alpha}.b(\ell\,\overline{5})(b(\ell\,\overline{8})(\ell\,\overline{7}))$$

Exercise 37. Prove that the closed long normal forms of type **tree** are in one-to-one correspondence with leaf-labelled binary trees.

9 Type inference

In Section 6, we introduced the simply-typed lambda calculus, and we discussed what it means for a term to be well-typed. We have also asked the question, for a given term, whether it is typable or not.

In this section, we will discuss an algorithm that decides, given a term, whether it is typable or not, and if the answer is yes, it also outputs a type for the term. Such an algorithm is known as a *type inference algorithm*.

A weaker kind of algorithm is a *type checking algorithm*. A type checking algorithm takes as its input a term with full type annotations, as well as the types of any free variables, and it decides whether the term is well-typed or not. Thus, a type checking algorithm does not infer any types; the type must be given to it as an input and the algorithm merely checks whether the type is legal.

Many compilers of programming languages include a type checker, and programs that are not well-typed are typically refused. The compilers of some programming languages, such as ML or Haskell, go one step further and include a type inference algorithm. This allows programmers to write programs with no or very few type annotations, and the compiler will figure out the types automatically. This makes the programmer's life much easier, especially in the case of higher-order languages, where types such as $((A \to B) \to C) \to D$ are not uncommon and would be very cumbersome to write down. However, in the event that type inference *fails*, it is not always easy for the compiler to issue a meaningful error message that can help the human programmer fix the problem. Often, at least a basic understanding of how the type inference algorithm works is necessary for programmers to understand these error messages.

9.1 Principal types

A simply-typed lambda term can have more than one possible type. Suppose that we have three basic types $\iota_1, \iota_2, \iota_3$ in our type system. Then the following are all valid typing judgments for the term $\lambda x.\lambda y.yx$:

$$
\begin{array}{l}
\vdash \lambda x^{\iota_1}.\lambda y^{\iota_1\to\iota_1}.yx : \iota_1 \to (\iota_1 \to \iota_1) \to \iota_1, \\
\vdash \lambda x^{\iota_2\to\iota_3}.\lambda y^{(\iota_2\to\iota_3)\to\iota_3}.yx : (\iota_2 \to \iota_3) \to ((\iota_2 \to \iota_3) \to \iota_3) \to \iota_3, \\
\vdash \lambda x^{\iota_1}.\lambda y^{\iota_1\to\iota_3}.yx : \iota_1 \to (\iota_1 \to \iota_3) \to \iota_3, \\
\vdash \lambda x^{\iota_1}.\lambda y^{\iota_1\to\iota_3\to\iota_2}.yx : \iota_1 \to (\iota_1 \to \iota_3 \to \iota_2) \to \iota_3 \to \iota_2, \\
\vdash \lambda x^{\iota_1}.\lambda y^{\iota_1\to\iota_1\to\iota_1}.yx : \iota_1 \to (\iota_1 \to \iota_1 \to \iota_1) \to \iota_1 \to \iota_1.
\end{array}
$$

What all these typing judgments have in common is that they are of the form

$$\vdash \lambda x^A.\lambda y^{A\to B}.yx : A \to (A \to B) \to B,$$

for certain types A and B. In fact, as we will see, *every* possible type of the term $\lambda x.\lambda y.yx$ is of this form. We also say that $A \to (A \to B) \to B$ is the *most general type* or the *principal type* of this term, where A and B are placeholders for arbitrary types.

The existence of a most general type is not a peculiarity of the term $\lambda xy.yx$, but it is true of the simply-typed lambda calculus in general: every typable term has a most general type. This statement is known as the *principal type property*.

We will see that our type inference algorithm not only calculates a possible type for a term, but in fact it calculates the most general type, if any type exists at all. In fact, we will prove the principal type property by closely examining the type inference algorithm.

9.2 Type templates and type substitutions

In order to formalize the notion of a most general type, we need to be able to speak of types with placeholders.

Definition. Suppose we are given an infinite set of *type variables*, which we denote by upper case letters X, Y, Z etc. A *type template* is a simple type, built from type variables and possibly basic types. Formally, type templates are given by the BNF

$$\text{Type templates:} \quad A, B ::= X \mid \iota \mid A \to B \mid A \times B \mid 1$$

Note that we use the same letters A, B to denote type templates that we previously used to denote types. In fact, from now on, we will simply regard types as special type templates that happen to contain no type variables.

The point of type variables is that they are placeholders (just like any other kind of variables). This means, we can replace type variables by arbitrary types, or even by type templates. A type substitution is just such a replacement.

Definition. A *type substitution* σ is a function from type variables to type templates. We often write $[X_1 \mapsto A_1, \ldots, X_n \mapsto A_n]$ for the substitution defined by $\sigma(X_i) = A_i$ for $i = 1 \ldots n$, and $\sigma(Y) = Y$ if $Y \notin \{X_1, \ldots, X_n\}$. If σ is a type substitution, and A is a type template, then we define $\bar{\sigma}A$, the *application* of σ to A, as follows by recursion on A:

$$\begin{array}{rcl}\bar{\sigma}X & = & \sigma X, \\ \bar{\sigma}\iota & = & \iota, \\ \bar{\sigma}(A \to B) & = & \bar{\sigma}A \to \bar{\sigma}B, \\ \bar{\sigma}(A \times B) & = & \bar{\sigma}A \times \bar{\sigma}B, \\ \bar{\sigma}1 & = & 1.\end{array}$$

In words, $\bar{\sigma}A$ is simply the same as A, except that all the type variables have been replaced according to σ. We are now in a position to formalize what it means for one type template to be more general than another.

Definition. Suppose A and B are type templates. We say that A is *more general* than B if there exists a type substitution σ such that $\bar{\sigma}A = B$.

In other words, we consider A to be more general than B if B can be obtained from A by a substitution. We also say that B is an *instance* of A. Examples:

- $X \to Y$ is more general than $X \to X$.
- $X \to X$ is more general than $\iota \to \iota$.
- $X \to X$ is more general than $(\iota \to \iota) \to (\iota \to \iota)$.
- Neither of $\iota \to \iota$ and $(\iota \to \iota) \to (\iota \to \iota)$ is more general than the other. We say that these types are *incomparable*.
- $X \to Y$ is more general than $W \to Z$, and vice versa. We say that $X \to Y$ and $W \to Z$ are *equally general*.

We can also speak of one substitution being more general than another:

Definition. If τ and ρ are type substitutions, we say that τ is more general than ρ if there exists a type substitution σ such that $\bar{\sigma} \circ \tau = \rho$.

9.3 Unifiers

We will be concerned with solving equations between type templates. The basic question is not very different from solving equations in arithmetic: given an equation between expressions, for instance $x + y = x^2$, is it possible to find values for x and y that make the equation true? The answer is yes in this case, for instance $x = 2, y = 2$ is one solution, and $x = 1, y = 0$ is another possible solution. We can even give the most general solution, which is $x =$ arbitrary, $y = x^2 - x$.

Similarly, for type templates, we might ask whether an equation such as

$$X \to (X \to Y) = (Y \to Z) \to W$$

has any solutions. The answer is yes, and one solution, for instance, is $X = \iota \to \iota$, $Y = \iota$, $Z = \iota$, $W = (\iota \to \iota) \to \iota$. But this is not the most general solution; the most general solution, in this case, is $Y =$ arbitrary, $Z =$ arbitrary, $X = Y \to Z$, $W = (Y \to Z) \to Y$.

We use substitutions to represent the solutions to such equations. For instance, the most general solution to the sample equation from the last paragraph is represented by the substitution

$$\sigma = [X \mapsto Y \to Z, W \mapsto (Y \to Z) \to Y].$$

If a substitution σ solves the equation $A = B$ in this way, then we also say that σ is a *unifier* of A and B.

To give another example, consider the equation

$$X \times (X \to Z) = (Z \to Y) \times Y.$$

This equation does not have any solution, because we would have to have both $X = Z \to Y$ and $Y = X \to Z$, which implies $X = Z \to (X \to Z)$, which is impossible to solve in simple types. We also say that $X \times (X \to Z)$ and $(Z \to Y) \times Y$ cannot be unified.

In general, we will be concerned with solving not just single equations, but systems of several equations. The formal definition of unifiers and most general unifiers is as follows:

Definition. Given two sequences of type templates $\bar{A} = A_1, \ldots, A_n$ and $\bar{B} = B_1, \ldots, B_n$, we say that a type substitution σ is a *unifier* of $\bar{A}$ and $\bar{B}$ if $\bar{\sigma}A_i = \bar{\sigma}B_i$, for all $i = 1 \ldots n$. Moreover, we say that σ is a *most general unifier* of $\bar{A}$ and $\bar{B}$ if it is a unifier, and if it is more general than any other unifier of $\bar{A}$ and $\bar{B}$.

9.4 The unification algorithm

Unification is the process of determining a most general unifier. More specifically, unification is an algorithm whose input are two sequences of type templates $\bar{A} = A_1, \ldots, A_n$ and $\bar{B} = B_1, \ldots, B_n$, and whose output is either "failure", if no unifier exists, or else a most general unifier σ. We call this algorithm mgu for "most general unifier", and we write $\mathrm{mgu}(\bar{A}; \bar{B})$ for the result of applying the algorithm to $\bar{A}$ and $\bar{B}$.

Before we state the algorithm, let us note that we only use finitely many type variables, namely, the ones that occur in $\bar{A}$ and $\bar{B}$. In particular, the substitutions generated by this algorithm are finite objects that can be represented and manipulated by a computer.

The algorithm for calculating $\mathrm{mgu}(\bar{A}; \bar{B})$ is as follows. By convention, the algorithm chooses the first applicable clause in the following list. Note that the algorithm is recursive.

1. $\mathrm{mgu}(X; X) = \mathrm{id}$, the identity substitution.
2. $\mathrm{mgu}(X; B) = [X \mapsto B]$, if X does not occur in B.
3. $\mathrm{mgu}(X; B)$ fails, if X occurs in B and $B \neq X$.
4. $\mathrm{mgu}(A; Y) = [Y \mapsto A]$, if Y does not occur in A.
5. $\mathrm{mgu}(A; Y)$ fails, if Y occurs in A and $A \neq Y$.
6. $\mathrm{mgu}(\iota; \iota) = \mathrm{id}$.
7. $\mathrm{mgu}(A_1 \to A_2; B_1 \to B_2) = \mathrm{mgu}(A_1, A_2; B_1, B_2)$.
8. $\mathrm{mgu}(A_1 \times A_2; B_1 \times B_2) = \mathrm{mgu}(A_1, A_2; B_1, B_2)$.
9. $\mathrm{mgu}(1; 1) = \mathrm{id}$.
10. $\mathrm{mgu}(A; B)$ fails, in all other cases.
11. $\mathrm{mgu}(A, \bar{A}; B, \bar{B}) = \bar{\tau} \circ \rho$, where $\rho = \mathrm{mgu}(\bar{A}; \bar{B})$ and $\tau = \mathrm{mgu}(\bar{\rho}A; \bar{\rho}B)$.

Note that clauses 1–10 calculate the most general unifier of two type templates, whereas clause 11 deals with lists of type templates. Clause 10 is a catch-all clause that fails if none of the earlier clauses apply. In particular, this clause causes the following to fail: $\mathrm{mgu}(A_1 \to A_2; B_1 \times B_2)$, $\mathrm{mgu}(A_1 \to A_2; \iota)$, etc.

Proposition 9.1. *If* $\mathrm{mgu}(\bar{A}; \bar{B}) = \sigma$, *then* σ *is a most general unifier of* $\bar{A}$ *and* $\bar{B}$. *If* $\mathrm{mgu}(\bar{A}; \bar{B})$ *fails, then* $\bar{A}$ *and* $\bar{B}$ *have no unifier.*

Proof. First, it is easy to prove by induction on the definition of mgu that if $\mathrm{mgu}(\bar{A}; \bar{B}) = \sigma$, then σ is a unifier of $\bar{A}$ and $\bar{B}$. This is evident in all cases except perhaps clause 11: but here, by induction hypothesis, $\bar{\rho}\bar{A} = \bar{\rho}\bar{B}$ and $\bar{\tau}(\bar{\rho}A) = \bar{\tau}(\bar{\rho}B)$, hence also $\bar{\tau}(\bar{\rho}(A, \bar{A})) = \bar{\tau}(\bar{\rho}(B, \bar{B}))$. Here we have used the evident notation of applying a substitution to a list of type templates.

Second, we prove that if $\bar{A}$ and $\bar{B}$ can be unified, then $\mathrm{mgu}(\bar{A}; \bar{B})$ returns a most general unifier. This is again proved by induction. For example, in clause 2, we have $\sigma = [X \mapsto B]$. Suppose τ is another unifier of X and B. Then $\bar{\tau}X = \bar{\tau}B$. We claim that $\bar{\tau} \circ \sigma = \tau$. But $\bar{\tau}(\sigma(X)) = \bar{\tau}(B) = \bar{\tau}(X) = \tau(X)$, whereas if $Y \neq X$, then $\bar{\tau}(\sigma(Y)) = \bar{\tau}(Y) = \tau(Y)$. Hence $\bar{\tau} \circ \sigma = \tau$, and it follows that σ is more general than τ. The clauses 1–10 all follow by similar arguments. For clause 11, suppose that $A, \bar{A}$ and $B, \bar{B}$ have some unifier σ'. Then σ' is also a unifier for $\bar{A}$ and $\bar{B}$, and thus the recursive call return a most general unifier ρ of $\bar{A}$ and $\bar{B}$. Since ρ is more general than σ', we have $\bar{\kappa} \circ \rho = \sigma'$ for some substitution κ. But then $\bar{\kappa}(\bar{\rho}A) = \bar{\sigma}'A = \bar{\sigma}'B = \bar{\kappa}(\bar{\rho}B)$, hence $\bar{\kappa}$ is a unifier for $\bar{\rho}A$ and $\bar{\rho}B$. By induction hypothesis, $\tau = \text{mgu}(\bar{\rho}A; \bar{\rho}B)$ exists and is a most general unifier for $\bar{\rho}A$ and $\bar{\rho}B$. It follows that τ is more general than $\bar{\kappa}$, thus $\bar{\kappa}' \circ \tau = \bar{\kappa}$, for some substitution κ'. Finally we need to show that $\sigma = \bar{\tau} \circ \rho$ is more general than σ'. But this follows because $\bar{\kappa}' \circ \sigma = \bar{\kappa}' \circ \bar{\tau} \circ \rho = \bar{\kappa} \circ \rho = \sigma'$. □

Remark 9.2. Proving that the algorithm mgu terminates is tricky. In particular, termination can't be proved by induction on the size of the arguments, because in the second recursive call in clause 11, the application of $\bar{\rho}$ may well increase the size of the arguments. To prove termination, note that each substitution σ generated by the algorithm is either the identity, or else it eliminates at least one variable. We can use this to prove termination by nested induction on the number of variables and on the size of the arguments. We leave the details for another time.

9.5 The type inference algorithm

Given the unification algorithm, type inference is now relatively easy. We formulate another algorithm, typeinfer, which takes a typing judgment $\Gamma \vdash M : B$ as its input (using templates instead of types, and not necessarily a *valid* typing judgment). The algorithm either outputs a most general substitution σ such that $\bar{\sigma}\Gamma \vdash M : \bar{\sigma}B$ is a valid typing judgment, or if no such σ exists, the algorithm fails.

In other words, the algorithm calculates the most general substitution that makes the given typing judgment valid. It is defined as follows:

1. $\text{typeinfer}(x_1{:}A_1, \ldots, x_n{:}A_n \vdash x_i : B) = \text{mgu}(A_i; B)$.

2. $\text{typeinfer}(\Gamma \vdash MN : B) = \bar{\tau} \circ \sigma$, where $\sigma = \text{typeinfer}(\Gamma \vdash M : X \to B)$, $\tau = \text{typeinfer}(\bar{\sigma}\Gamma \vdash N : \bar{\sigma}X)$, for a fresh type variable X.

3. $\text{typeinfer}(\Gamma \vdash \lambda x^A.M : B) = \bar{\tau} \circ \sigma$, where $\sigma = \text{mgu}(B; A \to X)$ and $\tau = \text{typeinfer}(\bar{\sigma}\Gamma, x{:}\bar{\sigma}A \vdash M : \bar{\sigma}X)$, for a fresh type variable X.

4. $\text{typeinfer}(\Gamma \vdash \langle M, N\rangle : A) = \bar{\rho} \circ \bar{\tau} \circ \sigma$, where $\sigma = \text{mgu}(A; X \times Y)$, $\tau = \text{typeinfer}(\bar{\sigma}\Gamma \vdash M : \bar{\sigma}X)$, and $\rho = \text{typeinfer}(\bar{\tau}\bar{\sigma}\Gamma \vdash N : \bar{\tau}\bar{\sigma}Y)$, for fresh type variables X and Y.

5. $\text{typeinfer}(\Gamma \vdash \pi_1 M : A) = \text{typeinfer}(\Gamma \vdash M : A \times Y)$, for a fresh type variable Y.

6. $\text{typeinfer}(\Gamma \vdash \pi_2 M : B) = \text{typeinfer}(\Gamma \vdash M : X \times B)$, for a fresh type variable X.

7. $\text{typeinfer}(\Gamma \vdash * : A) = \text{mgu}(A; 1)$.

Strictly speaking, the algorithm is non-deterministic, because some of the clauses involve choosing one or more fresh type variables, and the choice is arbitrary. However, the choice is not essential, since we may regard all fresh type variables are equivalent. Here, a type variable is called "fresh" if it has never been used.

Note that the algorithm typeinfer can fail; this happens if and only if the call to mgu fails in steps 1, 3, 4, or 7.

Also note that the algorithm obviously always terminates; this follows by induction on M, since each recursive call only uses a smaller term M.

Proposition 9.3. *If there exists a substitution σ such that $\bar{\sigma}\Gamma \vdash M : \bar{\sigma}B$ is a valid typing judgment, then* typeinfer$(\Gamma \vdash M : B)$ *will return a most general such substitution. Otherwise, the algorithm will fail.*

Proof. The proof is similar to that of Proposition 9.1. □

Finally, the question "is M typable" can be answered by choosing distinct type variables $X_1, \ldots, X_n, Y$ and applying the algorithm typeinfer to the typing judgment $x_1{:}X_1, \ldots, x_n{:}X_n \vdash M : Y$. Note that if the algorithm succeeds and returns a substitution σ, then σY is the most general type of M, and the free variables have types $x_1{:}\sigma X_1, \ldots, x_n{:}\sigma X_n$.

10 Denotational semantics

We introduced the lambda calculus as the "theory of functions". But so far, we have only spoken of functions in abstract terms. Do lambda terms correspond to any *actual* functions, such as, functions in set theory? And what about the notions of β- and η-equivalence? We intuitively accepted these concepts as expressing truths about the equality of functions. But do these properties really hold of real functions? Are there other properties that functions have that that are not captured by $\beta\eta$-equivalence?

The word "semantics" comes from the Greek word for "meaning". *Denotational semantics* means to give meaning to a language by interpreting its terms as mathematical objects. This is done by describing a function that maps syntactic objects (e.g., types, terms) to semantic objects (e.g., sets, elements). This function is called an *interpretation* or *meaning function*, and we usually denote it by $[\![-]\!]$. Thus, if M is a term, we will usually write $[\![M]\!]$ for the meaning of M under a given interpretation.

Any good denotational semantics should be *compositional*, which means, the interpretation of a term should be given in terms of the interpretations of its subterms. Thus, for example, $[\![MN]\!]$ should be a function of $[\![M]\!]$ and $[\![N]\!]$.

Suppose that we have an axiomatic notion of equality $\simeq$ on terms (for instance, $\beta\eta$-equivalence in the case of the lambda calculus). With respect to a particular class of interpretations, *soundness* is the property

$$M \simeq N \quad \Rightarrow \quad [\![M]\!] = [\![N]\!] \text{ for all interpretations in the class.}$$

Completeness is the property

$$[\![M]\!] = [\![N]\!] \text{ for all interpretations in the class} \quad \Rightarrow \quad M \simeq N.$$

Depending on our viewpoint, we will either say the axioms are sound (with respect to a given interpretation), or the interpretation is sound (with respect to a given set of axioms). Similarly for completeness. Soundness expresses the fact that our axioms (e.g., β or η) are true with respect to the given interpretation. Completeness expresses the fact that our axioms are sufficient.

10.1 Set-theoretic interpretation

The simply-typed lambda calculus can be given a straightforward set-theoretic interpretation as follows. We map types to sets and typing judgments to functions. For each basic type ι, assume that we have chosen a non-empty set S_ι. We can then associate a set $[\![A]\!]$ to each type A recursively:

$$\begin{array}{lcl} [\![\iota]\!] & = & S_\iota \\ [\![A \to B]\!] & = & [\![B]\!]^{[\![A]\!]} \\ [\![A \times B]\!] & = & [\![A]\!] \times [\![B]\!] \\ [\![1]\!] & = & \{*\} \end{array}$$

Here, for two sets X, Y, we write Y^X for the set of all functions from X to Y, i.e., $Y^X = \{f \mid f : X \to Y\}$. Of course, $X \times Y$ denotes the usual cartesian product of sets, and $\{*\}$ is some singleton set.

We can now interpret lambda terms, or more precisely, typing judgments, as certain functions. Intuitively, we already know which function a typing judgment corresponds to. For instance, the typing judgment $x{:}A, f{:}A \to B \vdash fx : B$ corresponds to the function that takes an element $x \in [\![A]\!]$ and an element $f \in [\![B]\!]^{[\![A]\!]}$, and that returns $f(x) \in [\![B]\!]$. In general, the interpretation of a typing judgment

$$x_1{:}A_1, \ldots, x_n{:}A_n \vdash M : B$$

will be a function

$$[\![A_1]\!] \times \ldots \times [\![A_n]\!] \to [\![B]\!].$$

Which particular function it is depends of course on the term M. For convenience, if $\Gamma = x_1{:}A_1, \ldots, x_n{:}A_n$ is a context, let us write $[\![\Gamma]\!] = [\![A_1]\!] \times \ldots \times [\![A_n]\!]$. We now define $[\![\Gamma \vdash M : B]\!]$ by recursion on M.

- If M is a variable, we define

  $$[\![x_1{:}A_1, \ldots, x_n{:}A_n \vdash x_i : A_i]\!] = \pi_i : [\![A_1]\!] \times \ldots \times [\![A_n]\!] \to [\![A_i]\!],$$

  where $\pi_i(a_1, \ldots, a_n) = a_i$.

- If $M = NP$ is an application, we recursively calculate

$$\begin{array}{rcl} f & = & [\![\Gamma \vdash N : A \to B]\!] : [\![\Gamma]\!] \to [\![B]\!]^{[\![A]\!]}, \\ g & = & [\![\Gamma \vdash P : A]\!] : [\![\Gamma]\!] \to [\![A]\!]. \end{array}$$

  We then define

$$[\![\Gamma \vdash NP : B]\!] = h : [\![\Gamma]\!] \to [\![B]\!]$$

  by $h(\bar{a}) = f(\bar{a})(g(\bar{a}))$, for all $\bar{a} \in [\![\Gamma]\!]$.

- If $M = \lambda x^A.N$ is an abstraction, we recursively calculate

$$f \ = \ [\![\Gamma, x{:}A \vdash N : B]\!] : [\![\Gamma]\!] \times [\![A]\!] \to [\![B]\!].$$

  We then define

$$[\![\Gamma \vdash \lambda x^A.N : A \to B]\!] = h : [\![\Gamma]\!] \to [\![B]\!]^{[\![A]\!]}$$

  by $h(\bar{a})(a) = f(\bar{a}, a)$, for all $\bar{a} \in [\![\Gamma]\!]$ and $a \in [\![A]\!]$.

- If $M = \langle N, P\rangle$ is an pair, we recursively calculate

$$\begin{array}{rcl} f & = & [\![\Gamma \vdash N : A]\!] : [\![\Gamma]\!] \to [\![A]\!], \\ g & = & [\![\Gamma \vdash P : B]\!] : [\![\Gamma]\!] \to [\![B]\!]. \end{array}$$

  We then define

$$[\![\Gamma \vdash \langle N, P\rangle : A \times B]\!] = h : [\![\Gamma]\!] \to [\![A]\!] \times [\![B]\!]$$

  by $h(\bar{a}) = (f(\bar{a}), g(\bar{a}))$, for all $\bar{a} \in [\![\Gamma]\!]$.

- If $M = \pi_i N$ is a projection (for $i = 1, 2$), we recursively calculate

$$f \ = \ [\![\Gamma \vdash N : B_1 \times B_2]\!] : [\![\Gamma]\!] \to [\![B_1]\!] \times [\![B_2]\!].$$

  We then define

$$[\![\Gamma \vdash \pi_i N : B_i]\!] = h : [\![\Gamma]\!] \to [\![B_i]\!]$$

  by $h(\bar{a}) = \pi_i(f(\bar{a}))$, for all $\bar{a} \in [\![\Gamma]\!]$. Here π_i in the meta-language denotes the set-theoretic function $\pi_i : [\![B_1]\!] \times [\![B_2]\!] \to [\![B_i]\!]$ given by $\pi_i(b_1, b_2) = b_i$.

- If $M = *$, we define

$$[\![\Gamma \vdash * : 1]\!] = h : [\![\Gamma]\!] \to \{*\}$$

  by $h(\bar{a}) = *$, for all $\bar{a} \in [\![\Gamma]\!]$.

To minimize notational inconvenience, we will occasionally abuse the notation and write $[\![M]\!]$ instead of $[\![\Gamma \vdash M : B]\!]$, thus pretending that terms are typing judgments. However, this is only an abbreviation, and it will be understood that the interpretation really depends on the typing judgment, and not just the term, even if we use the abbreviated notation.

We also refer to an interpretation as a *model*.

10.2 Soundness

Lemma 10.1 (Context change). *The interpretation behaves as expected under reordering of contexts and under the addition of dummy variables to contexts. More precisely, if $\sigma : \{1,\ldots,n\} \to \{1,\ldots,m\}$ is an injective map, and if the free variables of M are among $x_{\sigma 1},\ldots,x_{\sigma n}$, then the interpretations of the two typing judgments,*

$$\begin{aligned} f &= [\![x_1{:}A_1,\ldots,x_m{:}A_m \vdash M : B]\!] : [\![A_1]\!] \times \ldots \times [\![A_m]\!] \to [\![B]\!], \\ g &= [\![x_{\sigma 1}{:}A_{\sigma 1},\ldots,x_{\sigma n}{:}A_{\sigma n} \vdash M : B]\!] : [\![A_{\sigma 1}]\!] \times \ldots \times [\![A_{\sigma n}]\!] \to [\![B]\!] \end{aligned}$$

are related as follows:

$$f(a_1,\ldots,a_m) = g(a_{\sigma 1},\ldots,a_{\sigma n}),$$

for all $a_1 \in [\![A_1]\!],\ldots,a_m \in [\![A_m]\!]$.

Proof. Easy, but tedious, induction on M. □

The significance of this lemma is that, to a certain extent, the context does not matter. Thus, if the free variables of M and N are contained in Γ as well as Γ', then we have

$$[\![\Gamma \vdash M : B]\!] = [\![\Gamma \vdash N : B]\!] \quad \text{iff} \quad [\![\Gamma' \vdash M : B]\!] = [\![\Gamma' \vdash N : B]\!].$$

Thus, whether M and N have equal denotations only depends on M and N, and not on Γ.

Lemma 10.2 (Substitution Lemma). *If*

$$\begin{aligned} &[\![\Gamma, x{:}A \vdash M : B]\!] = f : [\![\Gamma]\!] \times [\![A]\!] \to [\![B]\!] \qquad \textit{and} \\ &[\![\Gamma \vdash N : A]\!] = g : [\![\Gamma]\!] \to [\![A]\!], \end{aligned}$$

then

$$[\![\Gamma \vdash M[N/x] : B]\!] = h : [\![\Gamma]\!] \to [\![B]\!],$$

where $h(\bar{a}) = f(\bar{a}, g(\bar{a}))$, for all $\bar{a} \in [\![\Gamma]\!]$.

Proof. Very easy, but very tedious, induction on M. □

Proposition 10.3 (Soundness). *The set-theoretic interpretation is sound for $\beta\eta$-reasoning. In other words,*

$$M =_{\beta\eta} N \quad \Rightarrow \quad [\![\Gamma \vdash M : B]\!] = [\![\Gamma \vdash N : B]\!].$$

Proof. Let us write $M \sim N$ if $[\![\Gamma \vdash M : B]\!] = [\![\Gamma \vdash N : B]\!]$. By the remark after Lemma 10.1, this notion is independent of Γ, and thus a well-defined relation on terms (as opposed to typing judgments). To prove soundness, we must show that

$M =_{\beta\eta} N$ implies $M \sim N$, for all M and N. It suffices to show that $\sim$ satisfies all the axioms of $\beta\eta$-equivalence.

The axioms (*refl*), (*symm*), and (*trans*) hold trivially. Similarly, all the (*cong*) and (ξ) rules hold, due to the fact that the meaning of composite terms was defined solely in terms of the meaning of their subterms. It remains to prove that each of the various (β) and (η) laws is satisfied (see page 49). We prove the rule ($\beta_{\to}$) as an example; the remaining rules are left as an exercise.

Assume Γ is a context such that $\Gamma, x{:}A \vdash M : B$ and $\Gamma \vdash N : A$. Let

$$\begin{array}{l} f = [\![\Gamma, x{:}A \vdash M : B]\!] : [\![\Gamma]\!] \times [\![A]\!] \to [\![B]\!], \\ g = [\![\Gamma \vdash N : A]\!] : [\![\Gamma]\!] \to [\![A]\!], \\ h = [\![\Gamma \vdash (\lambda x^A.M) : A \to B]\!] : [\![\Gamma]\!] \to [\![B]\!]^{[\![A]\!]}, \\ k = [\![\Gamma \vdash (\lambda x^A.M)N : B]\!] : [\![\Gamma]\!] \to [\![B]\!], \\ l = [\![\Gamma \vdash M[N/x] : B]\!] : [\![\Gamma]\!] \to [\![B]\!]. \end{array}$$

We must show $k = l$. By definition, we have $k(\bar{a}) = h(\bar{a})(g(\bar{a})) = f(\bar{a}, g(\bar{a}))$. On the other hand, $l(\bar{a}) = f(\bar{a}, g(\bar{a}))$ by the substitution lemma. □

Note that the proof of soundness amounts to a simple calculation; while there are many details to attend to, no particularly interesting new idea is required. This is typical of soundness proofs in general. Completeness, on the other hand, is usually much more difficult to prove and often requires clever ideas.

10.3 Completeness

We cite two completeness theorems for the set-theoretic interpretation. The first one is for the class of all models with finite base type. The second one is for the single model with one countably infinite base type.

Theorem 10.4 (Completeness, Plotkin, 1973). *The class of set-theoretic models with finite base types is complete for the lambda-$\beta\eta$ calculus.*

Recall that completeness for a class of models means that if $[\![M]\!] = [\![N]\!]$ holds in *all* models of the given class, then $M =_{\beta\eta} N$. This is not the same as completeness for each individual model in the class.

Note that, for each *fixed* choice of finite sets as the interpretations of the base types, there are some lambda terms such that $[\![M]\!] = [\![N]\!]$ but $M \neq_{\beta\eta} N$. For instance, consider terms of type $(\iota \to \iota) \to \iota \to \iota$. There are infinitely many $\beta\eta$-distinct terms of this type, namely, the Church numerals. On the other hand, if S_ι is a finite set, then $[\![(\iota \to \iota) \to \iota \to \iota]\!]$ is also a finite set. Since a finite set cannot have infinitely many distinct elements, there must necessarily be two distinct Church numerals M, N such that $[\![M]\!] = [\![N]\!]$.

Plotkin's completeness theorem, on the other hand, shows that whenever M and N are distinct lambda terms, then there exist *some* set-theoretic model with finite base types in which M and N are different.

The second completeness theorem is for a *single* model, namely the one where S_ι is a countably infinite set.

Theorem 10.5 (Completeness, Friedman, 1975). *The set-theoretic model with base type equal to $\mathbb{N}$, the set of natural numbers, is complete for the lambda-$\beta\eta$ calculus.*

We omit the proofs.

11 The language PCF

PCF stands for "programming with computable functions". The language PCF is an extension of the simply-typed lambda calculus with booleans, natural numbers, and recursion. It was first introduced by Dana Scott as a simple programming language on which to try out techniques for reasoning about programs. Although PCF is not intended as a "real world" programming language, many real programming languages can be regarded as (syntactic variants of) extensions of PCF, and many of the reasoning techniques developed for PCF also apply to more complicated languages.

PCF is a "programming language", not just a "calculus". By this we mean, PCF is equipped with a specific evaluation order, or rules that determine precisely how terms are to be evaluated. We follow the slogan:

Programming language = syntax + evaluation rules.

After introducing the syntax of PCF, we will look at three different equivalence relations on terms.

- *Axiomatic equivalence* $=_{\mathrm{ax}}$ will be given by axioms in the spirit of $\beta\eta$-equivalence.
- *Operational equivalence* $=_{\mathrm{op}}$ will be defined in terms of the operational behavior of terms. Two terms are operationally equivalent if one can be substituted for the other in any context without changing the behavior of a program.
- *Denotational equivalence* $=_{\mathrm{den}}$ is defined via a denotational semantics.

We will develop methods for reasoning about these equivalences, and thus for reasoning about programs. We will also investigate how the three equivalences are related to each other.

11.1 Syntax and typing rules

PCF types are simple types over two base types **bool** and **nat**.

$$A, B ::= \mathbf{bool} \mid \mathbf{nat} \mid A \to B \mid A \times B \mid 1$$

$$(true)\quad \frac{}{\Gamma \vdash \mathbf{T} : \mathbf{bool}} \qquad (pred)\quad \frac{\Gamma \vdash M : \mathbf{nat}}{\Gamma \vdash \mathbf{pred}\,(M) : \mathbf{nat}}$$

$$(false)\quad \frac{}{\Gamma \vdash \mathbf{F} : \mathbf{bool}} \qquad (iszero)\quad \frac{\Gamma \vdash M : \mathbf{nat}}{\Gamma \vdash \mathbf{iszero}\,(M) : \mathbf{bool}}$$

$$(zero)\quad \frac{}{\Gamma \vdash \mathbf{zero} : \mathbf{nat}} \qquad (fix)\quad \frac{\Gamma \vdash M : A \to A}{\Gamma \vdash \mathbf{Y}(M) : A}$$

$$(succ)\quad \frac{\Gamma \vdash M : \mathbf{nat}}{\Gamma \vdash \mathbf{succ}\,(M) : \mathbf{nat}}$$

$$(if)\quad \frac{\Gamma \vdash M : \mathbf{bool} \qquad \Gamma \vdash N : A \qquad \Gamma \vdash P : A}{\Gamma \vdash \mathbf{if}\ M\ \mathbf{then}\ N\ \mathbf{else}\ P : A}$$

Table 7: Typing rules for PCF

The raw terms of PCF are those of the simply-typed lambda calculus, together with some additional constructs that deal with booleans, natural numbers, and recursion.

$$\begin{array}{rcl} M,N,P & ::= & x \mid MN \mid \lambda x^A.M \mid \langle M,N\rangle \mid \pi_1 M \mid \pi_2 M \mid * \\ & & \mid \mathbf{T} \mid \mathbf{F} \mid \mathbf{zero} \mid \mathbf{succ}\,(M) \mid \mathbf{pred}\,(M) \\ & & \mid \mathbf{iszero}\,(M) \mid \mathbf{if}\ M\ \mathbf{then}\ N\ \mathbf{else}\ P \mid \mathbf{Y}(M) \end{array}$$

The intended meaning of these terms is the same as that of the corresponding terms we used to program in the untyped lambda calculus: $\mathbf{T}$ and $\mathbf{F}$ are the boolean constants, **zero** is the constant zero, **succ** and **pred** are the successor and predecessor functions, **iszero** tests whether a given number is equal to zero, **if** M **then** N **else** P is a conditional, and $\mathbf{Y}(M)$ is a fixed point of M.

The typing rules for PCF are the same as the typing rules for the simply-typed lambda calculus, shown in Table 4, plus the additional typing rules shown in Table 7.

11.2 Axiomatic equivalence

The axiomatic equivalence of PCF is based on the $\beta\eta$-equivalence of the simply-typed lambda calculus. The relation $=_{\text{ax}}$ is the least relation given by the following:

- All the β- and η-axioms of the simply-typed lambda calculus, as shown on page 49.
- One congruence or ξ-rule for each term constructor. This means, for instance

$$\frac{M =_{\text{ax}} M' \qquad N =_{\text{ax}} N' \qquad P =_{\text{ax}} P'}{\mathbf{if}\ M\ \mathbf{then}\ N\ \mathbf{else}\ P =_{\text{ax}} \mathbf{if}\ M'\ \mathbf{then}\ N'\ \mathbf{else}\ P'},$$

$$\begin{array}{rcl}
\mathbf{pred}\,(\mathbf{zero}\,) & = & \mathbf{zero} \\
\mathbf{pred}\,(\mathbf{succ}\,(\underline{n})) & = & \underline{n} \\
\mathbf{iszero}\,(\mathbf{zero}\,) & = & \mathbf{T} \\
\mathbf{iszero}\,(\mathbf{succ}\,(\underline{n})) & = & \mathbf{F} \\
\mathbf{if}\ \mathbf{T}\ \mathbf{then}\ N\ \mathbf{else}\ P & = & N \\
\mathbf{if}\ \mathbf{F}\ \mathbf{then}\ N\ \mathbf{else}\ P & = & P \\
\mathbf{Y}(M) & = & M(\mathbf{Y}(M))
\end{array}$$

Table 8: Axiomatic equivalence for PCF

and similar for all the other term constructors.

- The additional axioms shown in Table 8. Here, $\underline{n}$ stands for a *numeral*, i.e., a term of the form **succ** (... (**succ** (**zero**)) ...).

11.3 Operational semantics

The operational semantics of PCF is commonly given in two different styles: the *small-step* or *shallow* style, and the *big-step* or *deep* style. We give the small-step semantics first, because it is closer to the notion of β-reduction that we considered for the simply-typed lambda calculus.

There are some important differences between an operational semantics, as we are going to give it here, and the notion of β-reduction in the simply-typed lambda calculus. Most importantly, the operational semantics is going to be *deterministic*, which means, each term can be reduced in at most one way. Thus, there will never be a choice between more than one redex. Or in other words, it will always be uniquely specified which redex to reduce next.

As a consequence of the previous paragraph, we will abandon many of the congruence rules, as well as the (ξ)-rule. We adopt the following informal conventions:

- never reduce the body of a lambda abstraction,
- never reduce the argument of a function (except for primitive functions such as **succ** and **pred**),
- never reduce the "then" or "else" part of an if-then-else statement,
- never reduce a term inside a pair.

Of course, the terms that these rules prevent from being reduced can nevertheless become subject to reduction later: the body of a lambda abstraction and the argument of a function can be reduced after a β-reduction causes the λ to disappear and the argument to be substituted in the body. The "then" or "else" parts of an if-then-else term can be reduced after the "if" part evaluates to true or false.

$$\frac{M \to N}{\mathbf{pred}\,(M) \to \mathbf{pred}\,(N)}$$

$$\frac{}{\mathbf{pred}\,(\mathbf{zero}\,) \to \mathbf{zero}}$$

$$\frac{}{\mathbf{pred}\,(\mathbf{succ}\,(V)) \to V}$$

$$\frac{M \to N}{\mathbf{iszero}\,(M) \to \mathbf{iszero}\,(N)}$$

$$\frac{}{\mathbf{iszero}\,(\mathbf{zero}\,) \to \mathbf{T}}$$

$$\frac{}{\mathbf{iszero}\,(\mathbf{succ}\,(V)) \to \mathbf{F}}$$

$$\frac{M \to N}{\mathbf{succ}\,(M) \to \mathbf{succ}\,(N)}$$

$$\frac{M \to N}{MP \to NP}$$

$$\frac{}{(\lambda x^A.M)N \to M[N/x]}$$

$$\frac{M \to M'}{\pi_i M \to \pi_i M'}$$

$$\frac{}{\pi_1\langle M, N\rangle \to M}$$

$$\frac{}{\pi_2\langle M, N\rangle \to N}$$

$$\frac{M : 1, \quad M \neq *}{M \to *}$$

$$\frac{M \to M'}{\mathbf{if}\ M\ \mathbf{then}\ N\ \mathbf{else}\ P \to \mathbf{if}\ M'\ \mathbf{then}\ N\ \mathbf{else}\ P}$$

$$\frac{}{\mathbf{if}\ \mathbf{T}\ \mathbf{then}\ N\ \mathbf{else}\ P \to N}$$

$$\frac{}{\mathbf{if}\ \mathbf{F}\ \mathbf{then}\ N\ \mathbf{else}\ P \to P}$$

$$\frac{}{\mathbf{Y}(M) \to M(\mathbf{Y}(M))}$$

Table 9: Small-step operational semantics of PCF

And the terms inside a pair can be reduced after the pair has been broken up by a projection.

An important technical notion is that of a *value*, which is a term that represents the result of a computation and cannot be reduced further. Values are given as follows:

$$\text{Values:} \quad V, W ::= \mathbf{T} \mid \mathbf{F} \mid \mathbf{zero} \mid \mathbf{succ}\,(V) \mid * \mid \langle M, N\rangle \mid \lambda x^A.M$$

The transition rules for the small-step operational semantics of PCF are shown in Table 9.

We write $M \to N$ if M reduces to N by these rules. We write $M \not\to$ if there does not exist N such that $M \to N$. The first two important technical properties of small-step reduction are summarized in the following lemma.

Lemma 11.1. *1.* Values are normal forms. *If V is a value, then $V \not\to$.*

2. Evaluation is deterministic. *If $M \to N$ and $M \to N'$, then $N \equiv N'$.*

Another important property is subject reduction: a well-typed term reduces only to another well-typed term of the same type.

Lemma 11.2 (Subject Reduction). *If $\Gamma \vdash M : A$ and $M \to N$, then $\Gamma \vdash N : A$.*

Next, we want to prove that the evaluation of a well-typed term does not get "stuck". If M is some term such that $M \not\to$, but M is not a value, then we regard this as an error, and we also write $M \to \mathbf{error}$. Examples of such terms are $\pi_1(\lambda x.M)$ and $\langle M, N\rangle P$. The following lemma shows that well-typed closed terms cannot lead to such errors.

Lemma 11.3 (Progress). *If M is a closed, well-typed term, then either M is a value, or else there exists N such that $M \to N$.*

The Progress Lemma is very important, because it implies that a well-typed term cannot "go wrong". It guarantees that a well-typed term will either evaluate to a value in finitely many steps, or else it will reduce infinitely and thus not terminate. But a well-typed term can never generate an error. In programming language terms, a term that type-checks at *compile-time* cannot generate an error at *run-time*.

To express this idea formally, let us write $M \to^* N$ in the usual way if M reduces to N in zero or more steps, and let us write $M \to^* \mathbf{error}$ if M reduces in zero or more steps to an error.

Proposition 11.4 (Safety). *If M is a closed, well-typed term, then $M \not\to^* \mathbf{error}$.*

Exercise 38. Prove Lemmas 11.1–11.3 and Proposition 11.4.

11.4 Big-step semantics

In the small-step semantics, if $M \to^* V$, we say that M *evaluates to* V. Note that by determinacy, for every M, there exists at most one V such that $M \to^* V$.

It is also possible to axiomatize the relation "M evaluates to V" directly. This is known as the big-step semantics. Here, we write $M \Downarrow V$ if M evaluates to V. The axioms for the big-step semantics are shown in Table 10.

The big-step semantics satisfies properties similar to those of the small-step semantics.

Lemma 11.5.

1. Values. *For all values V, we have $V \Downarrow V$.*
2. Determinacy. *If $M \Downarrow V$ and $M \Downarrow V'$, then $V \equiv V'$.*
3. Subject Reduction. *If $\Gamma \vdash M : A$ and $M \Downarrow V$, then $\Gamma \vdash V : A$.*

The analogues of the Progress and Safety properties cannot be as easily stated for big-step reduction, because we cannot easily talk about a single reduction step or about infinite reduction sequences. However, some comfort can be taken in the fact that the big-step semantics and small-step semantics coincide:

Proposition 11.6. $M \to^* V$ *iff* $M \Downarrow V$.

$$\frac{}{\mathbf{T} \Downarrow \mathbf{T}}$$

$$\frac{}{\mathbf{F} \Downarrow \mathbf{F}}$$

$$\frac{}{\mathbf{zero} \Downarrow \mathbf{zero}}$$

$$\frac{}{\langle M, N \rangle \Downarrow \langle M, N \rangle}$$

$$\frac{}{\lambda x^A.M \Downarrow \lambda x^A.M}$$

$$\frac{M \Downarrow \mathbf{zero}}{\mathbf{pred}\,(M) \Downarrow \mathbf{zero}}$$

$$\frac{M \Downarrow \mathbf{succ}\,(V)}{\mathbf{pred}\,(M) \Downarrow V}$$

$$\frac{M \Downarrow \mathbf{zero}}{\mathbf{iszero}\,(M) \Downarrow \mathbf{T}}$$

$$\frac{M \Downarrow \mathbf{succ}\,(V)}{\mathbf{iszero}\,(M) \Downarrow \mathbf{F}}$$

$$\frac{M \Downarrow V}{\mathbf{succ}\,(M) \Downarrow \mathbf{succ}\,(V)}$$

$$\frac{M \Downarrow \lambda x^A.M' \qquad M'[N/x] \Downarrow V}{MN \Downarrow V}$$

$$\frac{M \Downarrow \langle M_1, M_2 \rangle \qquad M_1 \Downarrow V}{\pi_1 M \Downarrow V}$$

$$\frac{M \Downarrow \langle M_1, M_2 \rangle \qquad M_2 \Downarrow V}{\pi_2 M \Downarrow V}$$

$$\frac{M : 1}{M \Downarrow *}$$

$$\frac{M \Downarrow \mathbf{T} \qquad N \Downarrow V}{\mathbf{if}\ M\ \mathbf{then}\ N\ \mathbf{else}\ P \Downarrow V}$$

$$\frac{M \Downarrow \mathbf{F} \qquad P \Downarrow V}{\mathbf{if}\ M\ \mathbf{then}\ N\ \mathbf{else}\ P \Downarrow V}$$

$$\frac{M(\mathbf{Y}(M)) \Downarrow V}{\mathbf{Y}(M) \Downarrow V}$$

Table 10: Big-step operational semantics of PCF

11.5 Operational equivalence

Informally, two terms M and N will be called operationally equivalent if M and N are interchangeable as part of any larger program, without changing the observable behavior of the program. This notion of equivalence is also often called observational equivalence, to emphasize the fact that it concentrates on observable properties of terms.

What is an observable behavior of a program? Normally, what we observe about a program is its output, such as the characters it prints to a terminal. Since any such characters can be converted in principle to natural numbers, we take the point of view that the observable behavior of a program is a natural number that it evaluates to. Similarly, if a program computes a boolean, we regard the boolean value as observable. However, we do not regard abstract values, such as functions, as being directly observable, on the grounds that a function cannot be observed until we supply it some arguments and observe the result.

Definition. An *observable type* is either **bool** or **nat**. A *result* is a closed value of observable type. Thus, a result is either **T**, **F**, or $\underline{n}$. A *program* is a closed term of observable type.

A *context* is a term with a hole, written $C[-]$. Formally, the class of contexts is defined by a BNF:

$$C[-] \quad ::= \quad [-] \mid x \mid C[-]N \mid MC[-] \mid \lambda x^A.C[-] \mid \ldots$$

and so on, extending through all the cases in the definition of a PCF term.

Well-typed contexts are defined in the same way as well-typed terms, where it is understood that the hole also has a type. The free variables of a context are defined in the same way as for terms. Moreover, we define the *captured variables* of a context to be those bound variables whose scope includes the hole. So for instance, in the context $(\lambda x.[-])(\lambda y.z)$, the variable x is captured, the variable z is free, and y is neither free nor captured.

If $C[-]$ is a context and M is a term of the appropriate type, we write $C[M]$ for the result of replacing the hole in the context $C[-]$ by M. Here, we do not α-rename any bound variables, so that we allow free variables of M to be captured by $C[-]$.

We are now ready to state the definition of operational equivalence.

Definition. Two terms M, N are *operationally equivalent*, in symbols $M =_{\text{op}} N$, if for all closed and closing context $C[-]$ of observable type and all values V,

$$C[M] \Downarrow V \Leftrightarrow C[N] \Downarrow V.$$

Here, by a *closing* context we mean that $C[-]$ should capture all the free variables of M and N. This is equivalent to requiring that $C[M]$ and $C[N]$ are closed terms of observable types, i.e., programs. Thus, two terms are equivalent if they can be used interchangeably in any program.

11.6 Operational approximation

As a refinement of operational equivalence, we can also define a notion of operational approximation: We say that M *operationally approximates* N, in symbols $M \sqsubseteq_{\text{op}} N$, if for all closed and closing contexts $C[-]$ of observable type and all values V,

$$C[M] \Downarrow V \Rightarrow C[N] \Downarrow V.$$

Note that this definition includes the case where $C[M]$ diverges, but $C[N]$ converges, for some N. This formalizes the notion that N is "more defined" than M. Clearly, we have $M =_{\text{op}} N$ iff $M \sqsubseteq_{\text{op}} N$ and $N \sqsubseteq_{\text{op}} M$. Thus, we get a partial order $\sqsubseteq_{\text{op}}$ on the set of all terms of a given type, modulo operational equivalence. Also, this partial order has a least element, namely if we let $\Omega = \mathbf{Y}(\lambda x.x)$, then $\Omega \sqsubseteq_{\text{op}} N$ for any term N of the appropriate type.

Note that, in general, $\sqsubseteq_{\text{op}}$ is not a complete partial order, due to missing limits of ω-chains.

11.7 Discussion of operational equivalence

Operational equivalence is a very useful concept for reasoning about programs, and particularly for reasoning about program fragments. If M and N are operationally equivalent, then we know that we can replace M by N in any program without affecting its behavior. For example, M could be a slow, but simple subroutine for sorting a list. The term N could be a replacement that runs much faster. If we can prove M and N to be operationally equivalent, then this means we can safely use the faster routine instead of the slower one.

Another example are compiler optimizations. Many compilers will try to optimize the code that they produce, to eliminate useless instructions, to avoid duplicate calculations, etc. Such an optimization often means replacing a piece of code M by another piece of code N, without necessarily knowing much about the context in which M is used. Such a replacement is safe if M and N are operationally equivalent.

On the other hand, operational equivalence is a somewhat problematic notion. The problem is that the concept is not stable under adding new language features. It can happen that two terms, M and N, are operationally equivalent, but when a new feature is added to the language, they become nonequivalent, *even if M and N do not use the new feature*. The reason is the operational equivalence is defined in terms of contexts. Adding new features to a language also means that there will be new contexts, and these new contexts might be able to distinguish M and N.

This can be a problem in practice. Certain compiler optimizations might be sound for a sequential language, but might become unsound if new language features are added. Code that used to be correct might suddenly become incorrect if used in a richer environment. For example, many programs and library functions in C assume that they are executed in a single-threaded environment. If this code is ported to a multi-threaded environment, it often turns out to be no longer correct, and in many cases it must be re-written from scratch.

11.8 Operational equivalence and parallel or

Let us now look at a concrete example in PCF. We say that a term **POR** implements the *parallel or* function if it has the following behavior:

$$\begin{array}{lcll} \mathbf{POR}\ \mathbf{T}P & \to & \mathbf{T}, & \text{for all } P \\ \mathbf{POR}\ N\mathbf{T} & \to & \mathbf{T}, & \text{for all } N \\ \mathbf{POR}\ \mathbf{F}\mathbf{F} & \to & \mathbf{F}. & \end{array}$$

Note that this in particular implies $\mathbf{POR}\ \mathbf{T}\Omega = \mathbf{T}$ and $\mathbf{POR}\ \Omega\mathbf{T} = \mathbf{T}$, where Ω is some divergent term. It should be clear why **POR** is called the "parallel" or: the only way to achieve such behavior is to evaluate both its arguments in parallel, and to stop as soon as one argument evaluates to $\mathbf{T}$ or both evaluate to $\mathbf{F}$.

Proposition 11.7. **POR** *is not definable in PCF.*

We do not give the proof of this fact, but the idea is relatively simple: one proves by induction that every PCF context $C[-,-]$ with two holes has the following property: either, there exists a term N such that $C[M,M'] = N$ for all M, M' (i.e., the context does not look at M, M' at all), or else, either $C[\Omega, M]$ diverges for all M, or $C[M,\Omega]$ diverges for all M. Here, again, Ω is some divergent term such as $\mathbf{Y}(\lambda x.x)$.

Although **POR** is not definable in PCF, we can define the following term, called the *POR-tester*:

$$\begin{array}{rl} \textbf{POR-test} = \lambda x.\textbf{if } x\mathbf{T}\Omega \textbf{ then} & \\ & \textbf{if } x\Omega\mathbf{T} \textbf{ then} \\ & \quad \textbf{if } x\mathbf{FF} \textbf{ then } \Omega \\ & \quad \textbf{else } \mathbf{T} \\ & \textbf{else } \Omega \\ \textbf{else } \Omega & \end{array}$$

The POR-tester has the property that **POR-test** $M = \mathbf{T}$ if M implements the parallel or function, and in all other cases **POR-test** M diverges. In particular, since parallel or is not definable in PCF, we have that **POR-test** M diverges, for all PCF terms M. Thus, when applied to any PCF term, **POR-test** behaves precisely as the function $\lambda x.\Omega$ does. One can make this into a rigorous argument that shows that **POR-test** and $\lambda x.\Omega$ are operationally equivalent:

$$\textbf{POR-test} =_{\text{op}} \lambda x.\Omega \qquad \text{(in PCF).}$$

Now, suppose we want to define an extension of PCF called *parallel PCF*. It is defined in exactly the same way as PCF, except that we add a new primitive function **POR**, and small-step reduction rules

$$\frac{M \to M' \qquad N \to N'}{\textbf{POR } MN \to \textbf{POR } M'N'}$$

$$\frac{}{\textbf{POR } \mathbf{T}N \to \mathbf{T}}$$

$$\frac{}{\textbf{POR } M\mathbf{T} \to \mathbf{T}}$$

$$\frac{}{\textbf{POR } \mathbf{FF} \to \mathbf{F}}$$

Parallel PCF enjoys many of the same properties as PCF, for instance, Lemmas 11.1–11.3 and Proposition 11.4 continue to hold for it.

But notice that

$$\textbf{POR-test} \neq_{\text{op}} \lambda x.\Omega \qquad \text{(in parallel PCF).}$$

This is because the context $C[-] = [-]\,\textbf{POR}$ distinguishes the two terms: clearly, $C[\textbf{POR-test}] \Downarrow \mathbf{T}$, whereas $C[\lambda x.\Omega]$ diverges.

12 Complete partial orders

12.1 Why are sets not enough, in general?

As we have seen in Section 10, the interpretation of types as plain sets is quite sufficient for the simply-typed lambda calculus. However, it is insufficient for a language such as PCF. Specifically, the problem is the fixed point operator $\mathbf{Y} : (A \to A) \to A$. It is clear that there are many functions $f : A \to A$ from a set A to itself that do not have a fixed point; thus, there is no chance we are going to find an interpretation for a fixed point operator in the simple set-theoretic model.

On the other hand, if A and B are types, there are generally many functions $f : [\![A]\!] \to [\![B]\!]$ in the set-theoretic model that are not definable by lambda terms. For instance, if $[\![A]\!]$ and $[\![B]\!]$ are infinite sets, then there are uncountably many functions $f : [\![A]\!] \to [\![B]\!]$; however, there are only countably many lambda terms, and thus there are necessarily going to be functions that are not the denotation of any lambda term.

The idea is to put additional structure on the sets that interpret types, and to require functions to preserve that structure. This is going to cut down the size of the function spaces, decreasing the "slack" between the functions definable in the lambda calculus and the functions that exist in the model, and simultaneously increasing the chances that additional structure, such as fixed point operators, might exist in the model.

Complete partial orders are one such structure that is commonly used for this purpose. The method is originally due to Dana Scott.

12.2 Complete partial orders

Definition. A *partially ordered set* or *poset* is a set X together with a binary relation $\sqsubseteq$ satisfying

- *reflexivity:* for all $x \in X$, $x \sqsubseteq x$,
- *antisymmetry:* for all $x, y \in X$, $x \sqsubseteq y$ and $y \sqsubseteq x$ implies $x = y$,
- *transitivity:* for all $x, y, z \in X$, $x \sqsubseteq y$ and $y \sqsubseteq z$ implies $x \sqsubseteq z$.

The concept of a partial order differs from a total order in that we do not require that for any x and y, either $x \sqsubseteq y$ or $y \sqsubseteq x$. Thus, in a partially ordered set it is permissible to have incomparable elements.

We can often visualize posets, particularly finite ones, by drawing their line diagrams as in Figure 4. In these diagrams, we put one circle for each element of X, and we draw an edge from x upward to y if $x \sqsubseteq y$ and there is no z with $x \sqsubseteq z \sqsubseteq y$. Such line diagrams are also known as *Hasse diagrams*.

The idea behind using a partial order to denote computational values is that $x \sqsubseteq y$ means that x is *less defined than* y. For instance, if a certain term diverges, then its denotation will be less defined than, or below that of a term that has a

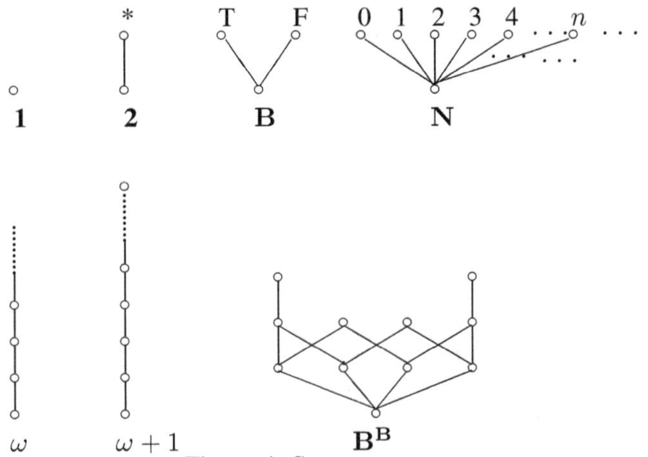

Figure 4: Some posets

definite value. Similarly, a function is more defined than another if it converges on more inputs.

Another important idea in using posets for modelling computational value is that of *approximation*. We can think of some infinite computational object (such as, an infinite stream), to be a limit of successive finite approximations (such as, longer and longer finite streams). Thus we also read $x \sqsubseteq y$ as *x approximates y*. A complete partial order is a poset in which every countable chain of increasing elements approximates something.

Definition. Let X be a poset and let $A \subseteq X$ be a subset. We say that $x \in X$ is an *upper bound* for A if $a \sqsubseteq x$ for all $a \in A$. We say that x is a *least upper bound* for A if x is an upper bound, and whenever y is also an upper bound, then $x \sqsubseteq y$.

Definition. An *ω-chain* in a poset X is a sequence of elements $x_0, x_1, x_2, \ldots$ such that

$$x_0 \sqsubseteq x_1 \sqsubseteq x_2 \sqsubseteq \ldots$$

Definition. A *complete partial order (cpo)* is a poset such that every ω-chain of elements has a least upper bound.

If $x_0, x_1, x_2, \ldots$ is an ω-chain of elements in a cpo, we write $\bigvee_{i \in \mathbb{N}} x_i$ for the least upper bound. We also call the least upper bound the *limit* of the ω-chain.

Not every poset is a cpo. In Figure 4, the poset labelled ω is not a cpo, because the evident ω-chain does not have a least upper bound (in fact, it has no upper bound at all). The other posets shown in Figure 4 are cpo's.

12.3 Properties of limits

Proposition 12.1. *1.* Monotonicity. *Suppose $\{x_i\}_i$ and $\{y_i\}_i$ are ω-chains in* *a cpo C, such that $x_i \sqsubseteq y_i$ for all i. Then*

$$\bigvee^{\uparrow}_{i} x_i \sqsubseteq \bigvee^{\uparrow}_{i} y_i.$$

2. Exchange. *Suppose $\{x_{ij}\}_{i,j\in\mathbb{N}}$ is a doubly monotone double sequence of elements of a cpo C, i.e., whenever $i \leqslant i'$ and $j \leqslant j'$, then $x_{ij} \sqsubseteq x_{i'j'}$. Then*

$$\bigvee^{\uparrow}_{i\in\mathbb{N}} \bigvee^{\uparrow}_{j\in\mathbb{N}} x_{ij} = \bigvee^{\uparrow}_{j\in\mathbb{N}} \bigvee^{\uparrow}_{i\in\mathbb{N}} x_{ij} = \bigvee^{\uparrow}_{k\in\mathbb{N}} x_{kk}.$$

In particular, all limits shown are well-defined.

Exercise 39. Prove Proposition 12.1.

12.4 Continuous functions

If we model data types as cpo's, it is natural to model algorithms as functions from cpo's to cpo's. These functions are subject to two constraints: they have to be monotone and continuous.

Definition. A function $f : C \to D$ between posets C and D is said to be *monotone* if for all $x, y \in C$,

$$x \sqsubseteq y \qquad \Rightarrow \qquad f(x) \sqsubseteq f(y).$$

A function $f : C \to D$ between cpo's C and D is said to be *continuous* if it is monotone and it preserves least upper bounds of ω-chains, i.e., for all ω-chains $\{x_i\}_{i\in\mathbb{N}}$ in C,

$$f(\bigvee^{\uparrow}_{i\in\mathbb{N}} x_i) = \bigvee^{\uparrow}_{i\in\mathbb{N}} f(x_i).$$

The intuitive explanation for the monotonicity requirement is that information is "positive": more information in the input cannot lead to less information in the output of an algorithm. The intuitive explanation for the continuity requirement is that any particular output of an algorithm can only depend on a finite amount of input.

12.5 Pointed cpo's and strict functions

Definition. A cpo is said to be *pointed* if it has a least element. The least element is usually denoted $\bot$ and pronounced "bottom". All cpo's shown in Figure 4 are pointed.

A continuous function between pointed cpo's is said to be *strict* if it preserves the bottom element.

12.6 Products and function spaces

If C and D are cpo's, then their *cartesian product* $C \times D$ is also a cpo, with the pointwise order given by $(x, y) \sqsubseteq (x', y')$ iff $x \sqsubseteq x'$ and $y \sqsubseteq y'$. Least upper bounds are also given pointwise, thus

$$\bigvee_i^{\uparrow}(x_i, y_i) = (\bigvee_i^{\uparrow} x_i, \bigvee_i^{\uparrow} y_i).$$

Proposition 12.2. *The first and second projections, $\pi_1 : C \times D \to C$ and $\pi_2 : C \times D \to D$, are continuous functions. Moreover, if $f : E \to C$ and $g : E \to D$ are continuous functions, then so is the function $h : E \to C \times D$ given by $h(z) = (f(z), g(z))$.*

If C and D are cpo's, then the set of continuous functions $f : C \to D$ forms a cpo, denoted D^C. The order is given pointwise: given two functions $f, g : C \to D$, we say that

$$f \sqsubseteq g \qquad \text{iff} \qquad \text{for all } x \in C,\ f(x) \sqsubseteq g(x).$$

Proposition 12.3. *The set D^C of continuous functions from C to D, together with the order just defined, is a complete partial order.*

Proof. Clearly the set D^C is partially ordered. What we must show is that least upper bounds of ω-chains exist. Given an ω-chain $f_0, f_1, \ldots$ in D^C, we define $g \in D^C$ to be the pointwise limit, i.e.,

$$g(x) = \bigvee_{i \in \mathbb{N}}^{\uparrow} f_i(x),$$

for all $x \in C$. Note that $\{f_i(x)\}_i$ does indeed form an ω-chain in C, so that g is a well-defined function. We claim that g is the least upper bound of $\{f_i\}_i$. First we need to show that g is indeed an element of D^C. To see that g is monotone, we use Proposition 12.1(1) and calculate, for any $x \sqsubseteq y \in C$,

$$g(x) = \bigvee_{i \in \mathbb{N}}^{\uparrow} f_i(x) \sqsubseteq \bigvee_{i \in \mathbb{N}}^{\uparrow} f_i(y) = g(y).$$

To see that g is continuous, we use Proposition 12.1(2) and calculate, for any ω-chain $x_0, x_1, \ldots$ in C,

$$g(\bigvee_j^{\uparrow} x_j) = \bigvee_i^{\uparrow}\bigvee_j^{\uparrow} f_i(x_j) = \bigvee_j^{\uparrow}\bigvee_i^{\uparrow} f_i(x_j) = \bigvee_j^{\uparrow} g(x_j).$$

Finally, we must show that g is the least upper bound of the $\{f_i\}_i$. Clearly, $f_i \sqsubseteq g$ for all i, so that g is an upper bound. Now suppose $h \in D^C$ is any other upper bound of $\{f_i\}$. Then for all x, $f_i(x) \sqsubseteq h(x)$. Since $g(x)$ was defined to be the least upper bound of $\{f_i(x)\}_i$, we then have $g(x) \sqsubseteq h(x)$. Since this holds for all x, we have $g \sqsubseteq h$. Thus g is indeed the least upper bound. □

Exercise 40. Recall the cpo $\mathbf{B}$ from Figure 4. The cpo $\mathbf{B}^{\mathbf{B}}$ is also shown in Figure 4. Its 11 elements correspond to the 11 continuous functions from $\mathbf{B}$ to $\mathbf{B}$. Label the elements of $\mathbf{B}^{\mathbf{B}}$ with the functions they correspond to.

Proposition 12.4. *The application function $D^C \times C \to D$, which maps (f, x) to $f(x)$, is continuous.*

Proposition 12.5. *Continuous functions can be continuously curried and uncurried. In other words, if $f : C \times D \to E$ is a continuous function, then $f^* : C \to E^D$, defined by $f^*(x)(y) = f(x, y)$, is well-defined and continuous. Conversely, if $g : C \to E^D$ is a continuous function, then $g_* : C \times D \to E$, defined by $g_*(x, y) = g(x)(y)$, is well-defined and continuous. Moreover, $(f^*)_* = f$ and $(g_*)^* = g$.*

12.7 The interpretation of the simply-typed lambda calculus in complete partial orders

The interpretation of the simply-typed lambda calculus in cpo's resembles the set-theoretic interpretation, except that types are interpreted by cpo's instead of sets, and typing judgments are interpreted as continuous functions.

For each basic type ι, assume that we have chosen a pointed cpo S_ι. We can then associate a pointed cpo $[\![A]\!]$ to each type A recursively:

$$\begin{array}{lcl} [\![\iota]\!] & = & S_\iota \\ [\![A \to B]\!] & = & [\![B]\!]^{[\![A]\!]} \\ [\![A \times B]\!] & = & [\![A]\!] \times [\![B]\!] \\ [\![1]\!] & = & \mathbf{1} \end{array}$$

Typing judgments are now interpreted as continuous functions

$$[\![A_1]\!] \times \ldots \times [\![A_n]\!] \to [\![B]\!]$$

in precisely the same way as they were defined for the set-theoretic interpretation. The only thing we need to check, at every step, is that the function defined is indeed continuous. For variables, this follows from the fact that projections of cartesian products are continuous (Proposition 12.2). For applications, we use the fact that the application function of cpo's is continuous (Proposition 12.4), and for lambda-abstractions, we use the fact that currying is a well-defined, continuous operation (Proposition 12.5). Finally, the continuity of the maps associated with products and projections follows from Proposition 12.2.

Proposition 12.6 (Soundness and Completeness). *The interpretation of the simply-typed lambda calculus in pointed cpo's is sound and complete with respect to the lambda-$\beta\eta$ calculus.*

12.8 Cpo's and fixed points

One of the reasons, mentioned in the introduction to this section, for using cpo's instead of sets for the interpretation of the simply-typed lambda calculus is that cpo's admit fixed point, and thus they can be used to interpret an extension of the lambda calculus with a fixed point operator.

Proposition 12.7. *Let C be a pointed cpo and let $f : C \to C$ be a continuous function. Then f has a least fixed point.*

Proof. Define $x_0 = \bot$ and $x_{i+1} = f(x_i)$, for all $i \in \mathbb{N}$. The resulting sequence $\{x_i\}_i$ is an ω-chain, because clearly $x_0 \sqsubseteq x_1$ (since x_0 is the least element), and if $x_i \sqsubseteq x_{i+1}$, then $f(x_i) \sqsubseteq f(x_{i+1})$ by monotonicity, hence $x_{i+1} \sqsubseteq x_{i+2}$. It follows by induction that $x_i \sqsubseteq x_{i+1}$. Let $x = \bigvee^{\uparrow}_i x_i$ be the limit of this ω-chain. Then using continuity of f, we have

$$f(x) = f(\bigvee^{\uparrow}_i x_i) = \bigvee^{\uparrow}_i f(x_i) = \bigvee^{\uparrow}_i x_{i+1} = x.$$

To prove that it is the least fixed point, let y be any other fixed point, i.e., let $f(y) = y$. We prove by induction that for all i, $x_i \sqsubseteq y$. For $i = 0$ this is trivial because $x_0 = \bot$. Assume $x_i \sqsubseteq y$, then $x_{i+1} = f(x_i) \sqsubseteq f(y) = y$. It follows that y is an upper bound for $\{x_i\}_i$. Since x is, by definition, the least upper bound, we have $x \sqsubseteq y$. Since y was arbitrary, x is below any fixed point, hence x is the least fixed point of f. □

If $f : C \to C$ is any continuous function, let us write $f^\dagger$ for its least fixed point. We claim that $f^\dagger$ depends continuously on f, i.e., that $\dagger : C^C \to C$ defines a continuous function.

Proposition 12.8. *The function $\dagger : C^C \to C$, which assigns to each continuous function $f \in C^C$ its least fixed point $f^\dagger \in C$, is continuous.*

Exercise 41. Prove Proposition 12.8.

Thus, if we add to the simply-typed lambda calculus a family of fixed point operators $Y_A : (A \to A) \to A$, the resulting extended lambda calculus can then be interpreted in cpo's by letting

$$[\![Y_A]\!] = \dagger : [\![A]\!]^{[\![A]\!]} \to [\![A]\!].$$

12.9 Example: Streams

Consider streams of characters from some alphabet A. Let $A^{\leqslant\omega}$ be the set of finite or infinite sequences of characters. We order A by the *prefix ordering*: if s and t are (finite or infinite) sequences, we say $s \sqsubseteq t$ if s is a prefix of t, i.e., if there exists a sequence s' such that $t = ss'$. Note that if $s \sqsubseteq t$ and s is an infinite sequence, then necessarily $s = t$, i.e., the infinite sequences are the maximal elements with respect to this order.

Exercise 42. Prove that the set $A^{\leqslant\omega}$ forms a cpo under the prefix ordering.

Exercise 43. Consider an automaton that reads characters from an input stream and writes characters to an output stream. For each input character read, it can write zero, one, or more output characters. Discuss how such an automaton gives rise to a continuous function from $A^{\leqslant\omega} \to A^{\leqslant\omega}$. In particular, explain the meaning of monotonicity and continuity in this context. Give some examples.

13 Denotational semantics of PCF

The denotational semantics of PCF is defined in terms of cpo's. It extends the cpo semantics of the simply-typed lambda calculus. Again, we assign a cpo $[\![A]\!]$ to each PCF type A, and a continuous function

$$[\![\Gamma \vdash M : B]\!] : [\![\Gamma]\!] \to [\![B]\!]$$

to every PCF typing judgment. The interpretation is defined in precisely the same way as for the simply-typed lambda calculus. The interpretation for the PCF-specific terms is shown in Table 11. Recall that $\mathbf{B}$ and $\mathbf{N}$ are the cpo's of lifted booleans and lifted natural numbers, respectively, as shown in Figure 4.

Definition. Two PCF terms M and N of equal types are denotationally equivalent, in symbols $M =_{\mathrm{den}} N$, if $[\![M]\!] = [\![N]\!]$. We also write $M \sqsubseteq_{\mathrm{den}} N$ if $[\![M]\!] \sqsubseteq [\![N]\!]$.

13.1 Soundness and adequacy

We have now defined the three notions of equivalence on terms: $=_{\mathrm{ax}}$, $=_{\mathrm{op}}$, and $=_{\mathrm{den}}$. In general, one does not expect the three equivalences to coincide. For example, any two divergent terms are operationally equivalent, but there is no reason why they should be axiomatically equivalent. Also, the POR-tester and the term $\lambda x.\Omega$ are operationally equivalent in PCF, but they are not denotationally equivalent (since a function representing POR clearly exists in the cpo semantics). For general terms M and N, one has the following property:

Theorem 13.1 (Soundness). *For PCF terms M and N, the following implications hold:*

$$M =_{\mathrm{ax}} N \qquad \Rightarrow \qquad M =_{\mathrm{den}} N \qquad \Rightarrow \qquad M =_{\mathrm{op}} N.$$

Soundness is a very useful property, because $M =_{\mathrm{ax}} N$ is in general easier to prove than $M =_{\mathrm{den}} N$, and $M =_{\mathrm{den}} N$ is in turns easier to prove than $M =_{\mathrm{op}} N$. Thus, soundness gives us a powerful proof method: to prove that two terms are operationally equivalent, it suffices to show that they are equivalent in the cpo semantics (if they are), or even that they are axiomatically equivalent.

Types:	$[\![\textbf{bool}]\!]$	$=$	$\mathbf{B}$
	$[\![\textbf{nat}]\!]$	$=$	$\mathbf{N}$
Terms:	$[\![\mathbf{T}]\!]$	$=$	$T \in \mathbf{B}$
	$[\![\mathbf{F}]\!]$	$=$	$F \in \mathbf{B}$
	$[\![\textbf{zero}]\!]$	$=$	$0 \in \mathbf{N}$
	$[\![\textbf{succ}\,(M)]\!]$	$=$	$\begin{cases} \bot & \text{if } [\![M]\!] = \bot, \\ n+1 & \text{if } [\![M]\!] = n \end{cases}$
	$[\![\textbf{pred}\,(M)]\!]$	$=$	$\begin{cases} \bot & \text{if } [\![M]\!] = \bot, \\ 0 & \text{if } [\![M]\!] = 0, \\ n & \text{if } [\![M]\!] = n+1 \end{cases}$
	$[\![\textbf{iszero}\,(M)]\!]$	$=$	$\begin{cases} \bot & \text{if } [\![M]\!] = \bot, \\ \mathbf{T} & \text{if } [\![M]\!] = 0, \\ \mathbf{F} & \text{if } [\![M]\!] = n+1 \end{cases}$
	$[\![\textbf{if } M \textbf{ then } N \textbf{ else } P]\!]$	$=$	$\begin{cases} \bot & \text{if } [\![M]\!] = \bot, \\ [\![N]\!] & \text{if } [\![M]\!] = \mathbf{F}, \\ [\![P]\!] & \text{if } [\![M]\!] = \mathbf{T}, \end{cases}$
	$[\![\mathbf{Y}(M)]\!]$	$=$	$[\![M]\!]^{\dagger}$

Table 11: Cpo semantics of PCF

As the above examples show, the converse implications are not in general true. However, the converse implications hold if the terms M and N are closed and of observable type, and if N is a value. This property is called computational adequacy. Recall that a program is a closed term of observable type, and a result is a closed value of observable type.

Theorem 13.2 (Computational Adequacy). *If M is a program and V is a result, then*

$$M =_{\text{ax}} V \qquad \iff \qquad M =_{\text{den}} V \qquad \iff \qquad M =_{\text{op}} V.$$

Proof. First note that the small-step semantics is contained in the axiomatic semantics, i.e., if $M \to N$, then $M =_{\text{ax}} N$. This is easily shown by induction on derivations of $M \to N$.

To prove the theorem, by soundness, it suffices to show that $M =_{\text{op}} V$ implies $M =_{\text{ax}} V$. So assume $M =_{\text{op}} V$. Since $V \Downarrow V$ and V is of observable type, it follows that $M \Downarrow V$. Therefore $M \to^* V$ by Proposition 11.6. But this already implies $M =_{\text{ax}} V$, and we are done. □

13.2 Full abstraction

We have already seen that the operational and denotational semantics do not coincide for PCF, i.e., there are some terms such that $M =_{\text{op}} N$ but $M \neq_{\text{den}} N$. Examples of such terms are **POR-test** and $\lambda x.\Omega$.

But of course, the particular denotational semantics that we gave to PCF is not the only possible denotational semantics. One can ask whether there is a better one. For instance, instead of cpo's, we could have used some other kind of mathematical space, such as a cpo with additional structure or properties, or some other kind of object altogether. The search for good denotational semantics is a subject of much research. The following terminology helps in defining precisely what is a "good" denotational semantics.

Definition. A denotational semantics is called *fully abstract* if for all terms M and N,

$$M =_{\text{den}} N \qquad \iff \qquad M =_{\text{op}} N.$$

If the denotational semantics involves a partial order (such as a cpo semantics), it is also called *order fully abstract* if

$$M \sqsubseteq_{\text{den}} N \qquad \iff \qquad M \sqsubseteq_{\text{op}} N.$$

The search for a fully abstract denotational semantics for PCF was an open problem for a very long time. Milner proved that there could be at most one such fully abstract model in a certain sense. This model has a syntactic description (essentially the elements of the model are PCF terms), but for a long time, no satisfactory semantic description was known. The problem has to do with sequentiality: a fully abstract model for PCF must be able to account for the fact that certain parallel constructs, such as parallel or, are not definable in PCF. Thus, the model should consist only of "sequential" functions. Berry and others developed a theory of "stable domain theory", which is based on cpo's with a additional properties intended to capture sequentiality. This research led to many interesting results, but the model still failed to be fully abstract.

Finally, in 1992, two competing teams of researchers, Abramsky, Jagadeesan and Malacaria, and Hyland and Ong, succeeded in giving a fully abstract semantics for PCF in terms of games and strategies. Games capture the interaction between a player and an opponent, or between a program and its environment. By considering certain kinds of "history-free" strategies, it is possible to capture the notion of sequentiality in just the right way to match PCF. In the last decade, game semantics has been extended to give fully abstract semantics to a variety of other programming languages, including, for instance, Algol-like languages.

Finally, it is interesting to note that the problem with "parallel or" is essentially the *only* obstacle to full abstraction for the cpo semantics. As soon as one adds "parallel or" to the language, the semantics becomes fully abstract.

Theorem 13.3. *The cpo semantics is fully abstract for parallel PCF.*

14 Acknowledgements

Thanks to Xiaoning Bian, Field Cady, Brendan Gillon, and Francisco Rios for reporting typos.

15 Bibliography

Here are some textbooks and other books on the lambda calculus. [1] is a standard reference handbook on the lambda calculus. [2]–[4] are textbooks on the lambda calculus. [5]–[7] are textbooks on the semantics of programming languages. Finally, [8]–[9] are textbooks on writing compilers for functional programming languages. They show how the lambda calculus can be useful in a more practical context.

[1] H. P. Barendregt. *The Lambda Calculus, its Syntax and Semantics.* North-Holland, 2nd edition, 1984.

[2] J.-Y. Girard, Y. Lafont, and P. Taylor. *Proofs and Types.* Cambridge University Press, 1989.

[3] J.-L. Krivine. *Lambda-Calculus, Types and Models.* Masson, 1993.

[4] G. E. Révész. *Lambda-Calculus, Combinators and Functional Programming.* Cambridge University Press, 1988.

[5] G. Winskel. *The Formal Semantics of Programming Languages. An Introduction.* MIT Press, London, 1993.

[6] J. C. Mitchell. *Foundations for Programming Languages.* MIT Press, London, 1996.

[7] C. A. Gunter. *Semantics of Programming Languages.* MIT Press, 1992.

[8] S. L. Peyton Jones. *The Implementation of Functional Programming Languages.* Prentice-Hall, 1987.

[9] A. W. Appel. *Compiling with Continuations.* Cambridge University Press, 1992.

www.ingramcontent.com/pod-product-compliance
Lightning Source LLC
Chambersburg PA
CBHW070948200526
45161CB00001BA/36
* 9 7 8 0 3 5 9 1 5 8 8 5 0 *